Polychlorinated Biphenyls

A Report Prepared by the

Committee on the Assessment of
Polychlorinated Biphenyls in the Environment

Environmental Studies Board
Commission on Natural Resources
National Research Council

NATIONAL ACADEMY OF SCIENCES
Washington, D.C. 1979

NOTICE: The project that is the subject of this report was approved by the Governing Board of the National Research Council, whose members are drawn from the Councils of the National Academy of Sciences, the National Academy of Engineering and the Institute of Medicine. The members of the Committee responsible for the report were chosen for their special competences and with regard for appropriate balance.

This report has been reviewed by a group other than the authors according to procedures approved by a Report Review Committee consisting of members of the National Academy of Sciences, the National Academy of Engineering, and the Institute of Medicine.

This study was supported by the Office of Research and Development, U.S. Environmental Protection Agency, Contract No. 68-01-3962.

Library of Congress Catalog Card Number 79-65443

International Standard Book Number 0-309-02885-X

Available from

Office of Publications
National Academy of Sciences
2101 Constitution Avenue, N.W.
Washington, D.C. 20418

Printed in the United States of America

COMMITTEE ON THE
ASSESSMENT OF
POLYCHLORINATED BIPHENYLS
IN THE ENVIRONMENT

ALFRED M. BEETON (Chairman), University of Michigan
JAMES R. ALLEN, University of Wisconsin
ANDERS W. ANDREN, University of Wisconsin
STANLEY J. CRISTOL, University of Colorado
ROBERT L. DURFEE, Versar, Inc.
A. MYRICK FREEMAN III, Bowdoin College
ROBERT L. METCALF, University of Illinois
IAN C.T. NISBET, Massachusetts Audubon Society
JOHN E. SANDERS, Barnard College
ANDERS SODERGREN, University of Lund
VLADIMIR ZITKO, Fisheries and Oceans Canada

SUELLEN W. PIRAGES, Staff Officer

ENVIRONMENTAL
STUDIES
BOARD

COMMISSION ON
NATURAL RESOURCES

CONTENTS

LIST OF ILLUSTRATIONS

Figures

LIST OF TABLES

Tables

PREFACE

Recognizing the hazards posed to human health and the environment by polychlorinated biphenyls (PCBs), the Office of Research and Development of the U.S. Environmental Protection Agency (EPA) requested that an assessment of the problem be conducted by the National Research Council. The Environmental Studies Board, given responsibility for the study, appointed the Committee on the Assessment of Polychlorinated Biphenyls in the Environment early in 1978. The charge to the Committee was to provide EPA with a detailed assessment of the nature of the problem and of alternatives for reducing the current substantial level of environmental contamination from PCBs.

Instead of contributing yet another extensive review of the fate of PCBs and the risks to health associated with them, the Committee set itself three objectives: (1) to develop a model of PCB distribution throughout the environment with estimates of current reservoir burdens, (2) to analyze the economic impact of various control options, and (3) using PCB data as the test case, to assess the effectiveness of the proposed EPA testing guidelines for evaluating toxic substances.

The Committee benefited greatly from discussions with individuals from academia, industry, and government agencies. We are particularly grateful to Frank Whitmore of Versar, Inc., for developing the model of PCB distribution. We also appreciate the assistance of several other persons at various stages of the preparation of the report: David Downer, Arthur Stern, Norbet Page (Office of Toxic Substances), Phill Taylor, Sam Conger (EPA Water Quality Analysis Branch-STORET), Wade Fowler (EPA Scientific Advisory Board), Steven Malkenson (EPA Office of Planning and Management), Nathan Karch (Council on Environmental Quality), Herman Feltz (U.S. Geological Survey), and Fred Kutz (EPA Office of Pesticides Program).

On behalf of the Committee, I would like to thank Suellen Pirages and Elizabeth Panos of the National Research Council for their contributions in preparing this report. Other staff members providing assistance included Judith Cummings, Nancy Fairfax, Raphael Kasper, Christina Olson, Connie Reges, William Robertson, and Robert Rooney.

Finally, I wish to express my gratitude to each member of the Committee for his efforts towards the successful completion of the study.

Alfred M. Beeton
Chairman, Committee on
the Assessment of Polychlorinated
Biphenyls in the Environment

INTRODUCTION

Thirty-six years after the introduction of polychlorinated biphenyls (PCBs) into the world market, scientists discovered that the compound was accumulating in tissues of fish taken from the Baltic Sea. Monitoring efforts within several countries subsequently disclosed extensive global environmental contamination from this group of organics, which comprises theoretically more than 200 individual isomers of chlorinated biphenyls. The problem of environmental contamination by PCBs has now been well documented. Comprehensive reviews have reported the existence of significant quantities of PCBs in atmosphere, soil, water, sediment, fish, wildlife, and even in samples of human blood and tissue (NIEHS 1972, Hutzinger et al. 1974, Kimbrough 1974, Durfee et al. 1976, Kornreich et al. 1976, Nisbet 1976, U.S. EPA 1976, WHO 1976, Roberts et al. 1978).

The term polychlorinated biphenyls refers to a class of organics produced by chlorination of a biphenyl and composed of 10 possible forms. These forms arise from a specified number of chlorine substitutions on the biphenyl molecule, and correspond to the chemical nomenclatures monochlorobiphenyl, dichlorobiphenyl, trichlorobiphenyl, etc. Several isomers for each PCB molecule are possible, the number depending on available substitution sites on each biphenyl portion (2-6, 2'-6'). However, not all possible isomers are likely to be formed during the manufacturing processes. In general, the most common ones are those that have either an equal number of chlorine atoms on both rings or a difference of only one chlorine atom between rings. More complete information on chemical identity and the manufacturing process is presented in Appendix D.

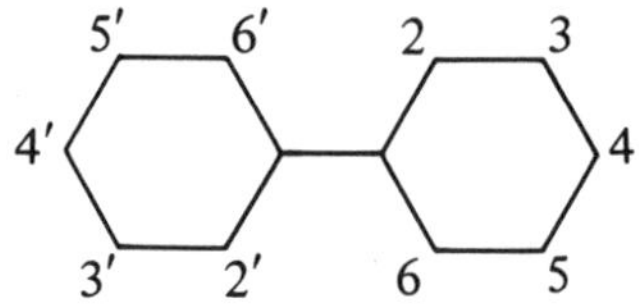

Commercial PCB mixtures are manufactured under a variety of trade names (Appendix D). The chlorine content of any product may vary from 18 to 79 percent and depends on the extent of chlorination during the manufacturing process or on the amount of isomeric mixing engaged in by individual producers. Each company has a specific system for identifying the chlorine content of its product. For example, Aroclor®1248, 1254 and 1260 indicate 48 percent, 54 percent and 60 percent chlorine, respectively; Clophen A60, Phenochlor DP6 and Kaneclor 600 designate that these products contain mixtures of hexachlorobiphenyls.

The only important U.S. producer of PCBs was Monsanto Industrial Chemicals Co., which had plants at Anniston, Alabama, where production of PCBs ended in 1970, and Sauget, Illinois, where production ceased in 1977. Sold under Monsanto's registered trademark of Aroclors, mixtures of PCBs had been used originally as a coolant/dielectric for transformers and capacitors, as heat transfer fluids, and as protective coatings for woods when low flammability was essential or desirable. Producers and users alike, apparently unaware of any potential hazards from exposure to PCBs, initially operated in accordance with earlier results of toxicity tests that indicated no effects (Penning 1930). The expansion of open-ended applications between 1930 and 1960, incorporating PCBs into such commodities as paints, inks, dedusting agents, and pesticides, led to the widespread dissemination of which we are now aware. By 1937, toxic effects were noted in occupationally exposed workers and threshold limit values were imposed at manufacturing sites.

The general pattern of release of PCBs to the environment changed significantly during the early 1970s. Until then, essentially no restrictions were imposed either on the use or on the disposal of PCBs. After evidence became available in 1969 and 1970 that chronic exposure could result in hazards to human health and the environment, Monsanto voluntarily banned sales of PCBs for "open-ended" and "nominally closed" applications (see Chapter 1, Tables 1.1 and 1.2), and the loss rate from industrial use was reduced through stringent control measures. However, significant reservoirs of mobile PCBs (those available for transport among environmental media and biota) still exist along with even larger, currently immobile reservoirs. The latter include those materials containing PCBs that are still in service and those deposited in landfills and dumps. The major factor affecting future release of PCBs from these sources will be regulatory action governing storage and disposal of the chemical.

The new Toxic Substances Control Act (TSCA), Public Law 94-469, specifically prohibits production of PCBs within the United States, regulates disposal of materials contaminated by PCBs, and restricts the use of any such materials already in service. The effect of these measures should be to eliminate further releases into the environment and, eventually, to reduce quantities existing in the environment. However, because of the extreme stability of PCBs, environmental levels will not be reduced substantially for many years, and the problem of dealing with existing reservoirs of mobile PCBs will remain.

Cleanup efforts have been initiated in those environments that are heavily contaminated, but complete removal may not be either technically or economically feasible. For this reason, the U.S. Environmental Protection Agency (EPA) commissioned the National Research Council to review the current dimensions of the PCB problem and to recommend future action.

The intent of this report is to provide EPA with an assessment of (1) the current state of the PCB problem, (2) the cost-effectiveness of selected options for controlling continued environmental pollution, and (3) the appropriateness of proposed guidelines for evaluating the potential hazard and persistence of chemicals with the hope of preventing future "PCB-like" problems. To accomplish its purpose this study has three primary components.

• Pertinent information on environmental aspects of PCBs was reviewed and summarized with emphasis on transport processes and distribution among environmental compartments. A model of the distribution of PCBs in the environment was constructed.

• An economic analysis was performed to determine, to the extent possible, the economic costs and environmental and health benefits that might be derived from implementing various options for controlling further PCB release and for reducing the present level of contamination. It should be noted that this analysis used the 1978 EPA published regulations for use and disposal of PCBs (U.S. EPA 1978a,b). New regulations are to be published in 1979 that may reflect different disposal requirements.

• Using EPA draft guidelines for testing potential toxicity of new chemicals, an assessment of PCB toxicity was made. An approach that might be used for identifying potential hazards prior to extensive manufacture and use of new chemicals was developed.

The study has a generic aspect in that the model of PCB distribution, the approach to quantifying persistence and toxicity, and the analysis of control options are applicable to studies of other persistent environmental pollutants.

REFERENCES

Durfee, R.L., G. Contos, F.C. Whitmore, J.D. Barden, E.E. Hackman III, and R.A. Westin (1976) PCBs in the United States--Industrial Use and Environmental Distribution. Report No. EPA 56016-16-005. Springfield, Va.: Versar, Inc.

Hutzinger, O., S. Safe, and V. Zitko (1974) The Chemistry of PCBs. Cleveland, Ohio: CRC Press.

Kimbrough, R.D. (1974) The toxicity of polychlorinated, polycyclic compounds and related chemicals. Pages 445-498, CRC Critical Reviews in Toxicology. Cleveland, Ohio: CRC Press.

Kornreich, M., B. Fuller, J. Dorigan, P. Walker, and L. Thomas (1976) Environmental Impact of

Polychlorinated Biphenyls. McLean, Va.: The Mitre Corporation.

National Institute of Environmental Health Sciences (1972) Conference on PCBs. Environ. Health Perspect. 1:1-85.

Nisbet, I.C.T. (1976) Criteria Document for PCBs. Report No. PB-225 397. Washington, D.C.: U.S. Department of Commerce.

Penning, C.H. (1930) Physical characteristics and commercial possibility of chlorinated diphenyl. Ind. Eng. Chem. 22:1180-1183.

Roberts, J.R., D.W. Rodgers, J.R. Bailey, and M.A. Rorke (1978) Polychlorinated Biphenyls: Biological Criteria for an Assessment of Their Effects on Environmental Quality. Associate Committee on Scientific Criteria for Environmental Quality. Ottawa, Canada: National Research Council of Canada.

U.S. Environmental Protection Agency (1976) Proceedings of National Conference on Polychlorinated Biphenyls, November 19-21, 1975, Chicago, Illinois. Report No. EPA-560/6-75-004. Washington, D.C.: U.S. Environmental Protection Agency.

U.S. Environmental Protection Agency (1978a) Polychlorinated biphenyls (PCBs): Disposal and marking. Federal Register 43(34):7149-7164.

U.S. Environmental Protection Agency (1978b) Polychlorinated biphenyls (PCBs): Manufacturing, processing, distribution in commerce, and use bans. Federal Register 43(110):24802-24818.

World Health Organization (1976) Polychlorinated Biphenyls and Terphenyls. Environmental Health Criteria 2. Geneva: World Health Organization.

FINDINGS AND RECOMMENDATIONS

Chapter 1. PCB Transport Throughout the Environment

1. A model of the distribution of PCBs throughout the U.S. environment estimates the total cumulative PCB use from domestic sources through 1975 as 6.1×10^8 kg, with 38 percent representing mono-, di-, and trichlorobiphenyl; 23 percent, tetrachlorobiphenyl; and 39 percent, penta- through decachlorobiphenyl.

2. From production and use data, losses of PCBs from production-use functions and the amounts entering the mobile environmental reservoir total 8.2×10^7 kg, distributed as follows:

	Amounts ($\times 10^7$ kg)
mono-, di-, trichlorobiphenyl:	3.1
tetrachlorobiphenyl:	1.8
penta- through decachlorobiphenyl:	3.3

3. The North Atlantic Ocean appears to be the dominant sink for PCBs, accounting for 50 to 80 percent of PCBs in the environment, and freshwater sediment is a major continental reservoir.

4. Environmental load estimates for the United States and the Atlantic Ocean are presented for the following compartments:

	Amounts (x 10^6 kg)
Atmosphere	0.018
Hydrosphere	
Fresh water	0.012 - 0.035
Freshwater sediment	1.400 - 7.100
Freshwater biota	0.030
Marine sediment	0.660 - 2.700
Marine water	6.000 - 66.00
Marine biota	0.030
Lithosphere	0.140 - 2.800

5. Biodegradation has been a minor factor in PCB destruction for those isomers of tetra- through decachlorobiphenyls.

Recommendations

1. Because the North Atlantic Ocean appears to be the ultimate sink, further studies are needed to firmly establish PCB concentrations and distribution in the marine environment.

2. Although modeling was useful in assessing large-scale distribution and fate of toxic substances, more adequate data bases are needed to develop transport models of environmental pollutants. For example, further research is needed to determine the processes of transfer between the environmental compartments.

3. EPA should continue to develop better analytical techniques, particularly those enabling assessment of the extent of PCB fractionation. Because of continuing analytical problems it is difficult to determine the extent of ecosystem contamination. For aquatic systems, the most effective indicator of both the degree of PCB pollution and the potential hazards to humans and wildlife is the PCB content of fish. To use this indicator effectively, factors such as age, species, and nutritional habits of the fish must be taken into account. Detailed information on individual isomers should also be reported together with the specific Aroclor designation.

4. Freshwater sediment is an important continental sink for PCBs, and special attention should be given to monitoring programs and to studies of PCB cycling from sediment through human food chains.

Chapter 2. Policy Assessment

1. Substantial differences exist in the costs of various options for the control of PCBs. Among the alternatives analyzed--which included EPA proposed regulations on disposal and use, options for dredging and disposal of contaminated river sediment, and U.S. Food and Drug Administration (FDA) regulations limiting permissible residues in fish destined for human consumption--average costs varied by as much as five orders of magnitude.

2. Existing EPA regulations on disposal and use are projected to result in the control of 260 x 10^6 kg of PCBs with average costs of up to $33/kg of PCBs controlled. Proposed EPA regulations would increase the total quantity of PCBs controlled by about 0.1 percent (beyond those controlled by existing regulations) with average costs becoming as high as $133,000/kg.

3. A case study of the Hudson River indicated that alternatives for dredging sediment could remove a total of about 178 x 10^3 kg of PCBs from potentially mobile river sediment at costs ranging from $110/kg to $3,150/kg.

4. It is estimated that up to 0.25 percent of the total PCBs released to the environment enter the human diet; at least 400 kg of PCBs would have to be removed from the mobile environmental reservoir in order to achieve a reduction of 1 kg of PCBs in food. This figure (400 kg) is derived using data from the Great Lakes, a closed system, and may be too low for more open systems. FDA limitations on PCB concentrations in commercially distributed fish could reduce PCB input to the American diet by about 64 kg at a cost of less than $125,000/kg. (A reduction of an additional 34 kg would cost less than $293,000/kg.)

5. Based on the figures used in the analysis in this report, FDA regulation of PCB limits in fish destined for human consumption appears a more cost-effective method of controlling dietary intake and human health effects than some of the most costly EPA proposed

regulations on use and disposal. Some dredging options may be preferable to either EPA or FDA regulations.

6. It is difficult to assess the benefits of controlling exposure to PCBs with currently available information. The analysis in this report conservatively estimates the benefits as \$30/kg to \$300/kg of PCBs removed from the environment or withheld from discharge, but notes that even these figures could be high or low by a factor of 100 or more.

Recommendations

1. Policymakers should use available data to calculate the incremental or average costs of PCB control options in conducting economic analyses of proposed regulations.

2. As greater control of PCBs is sought, regulatory efforts should be concentrated on those options which are most cost effective, that is those with the lowest average or incremental costs per comparable effect or benefit.

3. Economic analyses should include cost data on all options--not just those under a single agency's regulatory authority. These analyses could then show--as does the analysis presented in this report--that some options beyond a particular agency's authority are substantially more cost effective than some that lie within the agency's jurisdiction. A comprehensive, coordinated government policy toward control of pollutants should reflect such findings.

Chapter 3. Guidelines for Evaluating the Potential Toxicity of Chemicals

1. Based on the original uses of PCBs, this chemical might have passed toxic substance evaluation as proposed under the Federal Insecticide, Fungicide, and Rodenticide Act (FIFRA), thus allowing its use on a large scale. Acute toxicity is minimal for all species except some aquatic invertebrates, and intended use patterns may not have warranted additional chronic tests.

2. Results from the proposed TSCA Reference Set tests would require continued testing with the Hazard

Evaluation Studies leading to detection of <u>some</u> PCB toxic effects.

3. Most health hazards resulting from PCB exposure other than in occupational settings are subtle (e.g., reduced reproductive capabilities, changes in behavioral patterns) and would not be detected with either the FIFRA or TSCA scheme.

4. Common laboratory animals do not always respond to PCB exposure in a manner similar to humans. No one test species exhibits "typical" human symptoms of PCB poisoning. Thus, choice of test species becomes an important factor in evaluating the potential toxicity of a substance.

Recommendations

1. Whenever chemical, physical, and environmental chemistry data indicate that a substance is both persistent and lipophilic, its development for large-scale use should be reassessed and alternatives considered.

2. Testing schemes should include models of anticipated release and movement through the environment identifying potential reservoirs, and tests should be performed within these specified environmental ecosystems.

3. Test guidelines should stress assessment of subtle health hazards, such as reproductive alterations and behavioral-learning impairment.

1

PCB TRANSPORT THROUGHOUT THE ENVIRONMENT

In considering PCB sources, mechanisms of transport through the environment, and mobile or immobile reservoirs, it is often useful to consider global cycles (Nisbet and Sarofim 1972). Models of these cycles can provide insights into the general behavior of compounds and can establish a framework for subsequent research. This chapter presents a review of PCB distribution in the major environmental compartments for the United States (i.e., atmosphere, hydrosphere, lithosphere) and the means of transport among and within them.

In any attempt to design a distribution model based on results from a number of different studies, several possible sources of error must be recognized. Lack of uniformity in sampling and analytical techniques is the major problem encountered. Sampling difficulties occur for measurements in air, water, and sediment. In air, the amount of PCBs in the vapor phase relative to that attached to particles still is uncertain, mainly because of problems in sampling techniques. Similar difficulties are encountered in efforts to distinguish between dissolved vs. particle-bound PCBs in water. Measurement of PCBs in sediment also presents problems. Most studies have not used existing, proven techniques for collecting sediment cores, but instead have analyzed "grab" samples. Grab sample analyses can result in overestimates or underestimates of PCB levels depending on the depth of sample taken, and may not reflect accurately the concentration of the sediment column.

The comparisons of measurements of PCBs in environmental samples have also been complicated by a lack of uniformity in analytical techniques. Because each commercial mixture contains a number of PCB isomers and thus exhibits a wide range of vapor pressures and solubility characteristics, the most common approach to quantifying PCB concentrations is to compare gas chromatographic records of the sample with records of Aroclor, Clophen, Phenoclor or Kanechlor standards. But this approach is not entirely appropriate, unless both the number of peaks recorded and the interpeak height ratios are identical for the sample and standard being compared (Webb and McCall 1973, Murphy and Rzeszutko 1977). Analytical fractionation techniques make it difficult to identify adequately all peaks on the gas chromatographic record, and it is quite possible that underestimates of the quantities of mono- and dichlorobiphenyls have resulted. Because of these problems in identifying the chemical components of a given mixture, PCB concentrations are often reported as "most closely resembling" or calculated as "similar to" a certain known Aroclor standard. Thus, it becomes difficult to discern whether or to what extent individual component fractionations actually occur within the environment.

The ability to understand and assess PCB movement throughout the environment can only remain speculative until the sampling and analytical problems are resolved; the following interpretations of data were made with these limitations in mind.

SOURCES OF CONTAMINATION

Estimates of General PCB Distribution Within the United States

During the period from 1930 to 1975, U.S. commercial sales of PCBs totalled nearly 570 x 10^6 kg from domestic sources and about 1.4 x 10^6 kg from imports (Durfee et al. 1976). As of 1975, more than half of this amount, or 340 x 10^6 kg, was estimated to be in service.

Distribution of the remainder was estimated using 1957 to 1974 production and sales figures supplied by Monsanto, information supplied by other industrial concerns, and analysis of PCB use patterns with respect to losses, waste disposal, and other aspects of environmental entry. The amount and location of PCBs are presented in Table 1.1.

Routes of Entry into the Environment

There is no substantial evidence suggesting that PCBs are produced in the environment, either from natural sources or from chemical transformation of other compounds. Therefore, all environmental contamination by PCBs is inferred to have resulted from the production and use of materials and equipment containing PCBs. It also can be inferred, with some certainty, that PCBs entering the United States through commercial imports or via transnational atmospheric transport have been inconsequential when compared to the amounts entering the U.S. environment via domestic routes.

Major domestic uses of PCBs are indicated in Table 1.2. The voluntary limitation of PCB production imposed by Monsanto in 1971 restricted sales to applications in closed electrical systems. The 1973 estimated imports of PCBs to the United States totalled only 1.0 percent of Monsanto's production for that year, suggesting further voluntary elimination of uses other than for closed systems. The major applications of imported PCBS at that time were for pattern waxes (decachlorobiphenyl) and replacement coolant fluids for mining machines. Further evidence of the decrease in PCB use lies in the

TABLE 1.1 Environmental Distribution of PCBs Not in Service as of 1975

	Amount ($\times 10^6$ kg)
Mobile PCBs in the Environment	68.2
Degraded or Incinerated	25
In Landfills or Equipment Dumps	130

SOURCE: Modified from Durfee et al. (1976).

TABLE 1.2 Domestic Uses of PCBs

Category	Type of Product	Percent of New Total Use
Closed Electrical Systems	Transformer, capacitors, other (minor) electrical insulating/cooling applications	61 until 1971; 100 after 1971
Nominally Closed Systems	Hydraulic fluids, heat transfer fluids, lubricants	13 until 1971; 0 after 1971
Open-End Applications	Plasticizers, surface coatings, ink and dye carriers, adhesives, pesticide extenders, carbonless copy paper, dyes	26 until 1971; 0 after 1971

fact that although Monsanto ceased all PCB production in 1977, large-scale increases in imports of these compounds did not occur.

Two important routes of entry into the environment have been losses during the process of manufacture and leakage from electrical equipment and other products containing PCBs. Losses from open-end and nominally closed systems in service undoubtedly have been large and have contributed to contamination of soil, water, and atmosphere.

Mobilization of PCBs located in landfills and freshwater sediment offers significant additional potential for contamination, but few data are available for estimating current or future rates of dispersion. In studies of New York landfills some volatilization of PCBs is noted, but movement of PCBs from these sources into groundwater reservoirs is negligible (Leis and Metry 1978, Hetling et al. 1978). Disposal of materials containing PCBs--as for example, through municipal incineration--appears to be one of many minor routes of entry, and application of PCB-contaminated oils as dedusting agents may mobilize relatively large quantities of PCBs formerly in service.

A summary estimate of the quantity of PCBs in landfills as of 1975 is shown in Table 1.3. The estimated rate of PCB entry into on-land disposal sites from all sources in 1975 was 5.5 x 10^6 kg/yr (Durfee et al. 1976). Assuming that this estimate is valid for the period 1973 through 1978, the total accumulation in landfills by 1978 can be estimated as 140 x 10^6 kg. However, this total may be an overestimate as disposal on land may have declined with the passage of TSCA and the attention given this aspect of the PCB problem. Some obsolete equipment has been stored instead of dumped in landfills, but the major reduction of PCBs deposited in land sites has resulted through a decrease in production wastes due to diminished domestic manufacture and use of PCBs.

In the future, the primary source of PCB entry to on-land disposal sites is likely to be the obsolescence of electrical equipment, because no regulations are anticipated that could change this situation. Thus, it is not unreasonable to suggest a maximum entry rate of about 5.5 x 10^6 kg/yr to landfills for at least 10 years and perhaps as many as 20, although by the year 2000 the rate should have decreased somewhat.

PCBs newly deposited in landfills should be much less mobile than those in old landfill sites, because of the Resource Conservation and Recovery Act (RCRA) requirements for better designed landfills and tighter controls on monitoring site deterioration and chemical leakage. Much of the new deposits will be contained sufficiently to prevent leakage or volatilization and new sites must be designed to reduce mobilization of toxic materials by leaching.

Sources of PCBs

Industrial and Municipal Effluents Recent estimates of industrial and municipal PCB discharges into aquatic systems have been determined through monitoring programs

TABLE 1.3 Reservoir of PCBs in Landfills as of 1975

Source	Estimated Amount (x 10^6 kg)
Production Wastes	50
Obsolete Equipment	36
Other	36
Total	122

SOURCE: Modified from Durfee et al. (1976).

initiated by EPA. The results indicate that the average PCB content of industrial effluents (five categories) is 0.68×10^3 kg/yr, and the PCB content of discharges from publicly owned wastewater treatment works (POTW) is 110×10^3 kg/yr (Versar, Inc. 1978).

The sampling efforts producing the industrial discharge data, on which these estimates were based, have been far from comprehensive, but they may provide results representative of the five industrial segments covered (i.e., timber processing, nonferrous metals, ore mining, pulp and paper manufacturing, and machinery and mechanical products). The estimate of PCB discharge from POTW is high by perhaps a factor of three and reflects a nonrandom sampling of 66 treatment plants; some of these were in the vicinity of known PCB operations (Versar, Inc. 1978).

Even if the estimated POTW discharge rate is high, the municipal sector apparently releases a significant amount of PCBs into the aquatic environment. The major source is believed to be the mobile environmental reservoir, because the municipal wastes are presumed to have passed from a use cycle to the environment before entering a wastewater stream. Leakage from equipment in use or being serviced accounts for the remainder of PCBs in POTW discharges.

If it is assumed that approximately 40×10^3 kg/yr of total PCBs are discharged from municipal sewage-treatment plants, the release rate of PCBs in sewage sludge that is released to the environment should total about 240×10^3 kg/yr based on an approximate 6:1 ratio for partitioning of PCBs between sludge and water effluent (Sargent et al. 1977). Approximately 85 to 90 percent of PCBs in the portion of sludge that is incinerated (on the order of 15 percent of total sludge) may be destroyed via combustion (Whitmore 1975); the remaining PCBs in sludge represent a potential threat to the environment as well as a substantial but perhaps temporary sink for PCBs that were previously environmentally mobile.

Paper Recycling The use of PCBs in carbonless copy paper from 1957 through 1971 has resulted in the contamination of paper products and foods by recycling practices and entry of contaminated effluents into the environment via paper-recycling facilities. Projections based on available data have indicated that both average concentrations and occurrence of "hot spots" are now decreasing rapidly and will continue to decrease (Carr et al. 1977). Some fraction of the PCBs in carbonless copy paper and recycled paper is also released to the atmosphere through incineration under conditions that do not destroy PCBs completely.

<u>Release of PCBs to Soil</u> In light of the diminishing use of PCBs in industry, the major routes of future entry into soil apparently will be from disposal sites, in-service equipment, and atmospheric transport. Disposal sites represent a sizable reservoir of PCBs, but little is known about the lifetime of these compounds in a site (for example, a landfill) or their effective rate of release. A recent study of the Hudson River area indicates that negligible groundwater contamination by PCBs has occurred in areas surrounding old landfills (Leis and Metry 1978, Hetling et al. 1978), but extrapolation from the local to the national scale is not possible.

Accidental and deliberate releases pose a potential threat of excessive contamination, but only locally. Diminished industrial use should result in a reduction in the number of spills. It is also hoped that application of PCBs or PCB-contaminated oils for dust control will be reduced through public awareness and regulatory actions; but if it is not, such applications will continue to have local impact on concentrations of PCBs in soil.

<u>Incineration</u> The PCB regulations proposed by EPA suggest that ultimately incineration will be the required method of disposal, but at this time it is not clear whether the necessary facilities will be available, or what conditions of burning time and temperature are needed to incinerate large quantities of contaminated wastes (U.S. EPA 1978). Research is in progress on the incineration conditions required to destroy PCBs, but the investigation is not yet complete. One problem needing immediate resolution concerns the effect of applying "combustion efficiency" criteria based strictly on combustion conditions (CO/CO_2 in exhaust) rather than on destruction of PCBs by the incineration process.

<u>Chemical Landfills</u> The situation is also unclear regarding availability and adequacy of chemical waste landfills for disposing of PCBs. At present, four such

sites exist in the United States and more are planned, but little information is available about the potential for loss of PCBs from landfills through long periods of residence. Another problem, affecting both incineration and disposal in landfills, is how to assign responsibility for spills or other accidental losses that may occur during transport of contaminated waste across state lines.

ATMOSPHERIC COMPARTMENT

Recently available data indicate that atmospheric transport is an important factor in PCB distribution on both local and global scales (Harvey and Steinhauer 1976a, Risebrough et al. 1976, Fuller et al. 1977, and Bidleman et al. 1978). Quantities of the compound detected in a variety of samples collected at remote areas (e.g., in water, snow, and soil) provide strong evidence for this conclusion, even though the details of atmospheric transport are not well understood. Current data do not permit accurate characterization of such factors as vapor/particulate partitioning, individual component fractionation, spatial and temporal variations in deposition, volatilization or degradation rates, or identification of ultimate sinks. Accurate measurement has been hampered, as stated earlier, mainly by technical problems in sampling and analytical procedures. But despite the problems, it has been possible to reach several important conclusions concerning atmospheric transport of PCBs.

Physical and Chemical Forms of Atmospheric PCBs

Current information about PCB vapor pressures suggests that PCBs should exist both in the vapor form and attached to particles in the atmosphere. Determination of PCB levels in air has indicated that more than 99 percent of the total atmospheric burden of PCBs can pass through a glass-fiber filter and thus must be collected on various adsorbent traps (e.g., silicone oil coated ceramic chips, polyurethane foam, or XAD-2 resin [Bidleman et al. 1978]). However, it cannot be determined whether PCBs are associated with particles too small to be captured by the glass-fiber filter or whether 99 percent of the total burden of PCBs actually exists as a vapor. It is thought that this collection method may be further hampered by vaporization of PCBs associated with particulates during the collection process. However, Harvey and Steinhauer (1974) state:

> ...the observation that only 1 percent or less of the total chlorinated hydrocarbon concentration in the sampled air is on the filter argues strongly against significant particle transport.

Murphy and Rzeszutko (1977) tried to resolve this phase-partitioning question by measuring individual Aroclors in both air and precipitation (Tables 1.4 and 1.5). Two assumptions were made. The first was that all Aroclor 1260 detected in air is associated with particulates; it is likely that this is a valid asssumption because the vapor pressure range for Aroclor 1260 compounds is on the order of 10^{-4} mm Hg to 10^{-5} mm Hg. The second assumption, based upon the magnitude of an effective Henry's Law Constant for Aroclor 1260, contended that all PCBs in rain result from particulate scavenging. This assumption might be in error because of the uncertainties involved in calculating and measuring Henry's Law Constant. The value of the partition coefficient calculated from Henry's Law Constant indicates that the concentrations of scavenged vapor in precipitation should be negligible (Mackay and Wolkoff 1973). Using their measured Aroclor partitioning and the above assumptions, Murphy and

TABLE 1.4 PCBs in Precipitation

Location (Number of Samples)	Average Concentration (Range) (ng/l)	Percent Dissolved
Norway, Rain (5)[1]	4.7 (1.2 - 7.6)	44
Norway, Snow (11)	4.2 (1.5 - 7.0)	56
Lake Michigan, Rain[2]	120[a] (190 ± 140)[b]	66
Lake Michigan, Snow (4)	210 ± 97	36

[a]Weighted mean concentration.
[b]Arithmetic mean.

SOURCE:
[1]Modified from Lunde et al. (1977).
[2]Modified from Murphy and Rzeszutko (1977).

TABLE 1.5 PCB Concentrations in Continental Air (10^{-9} g/m^3)

Location and Date	No. Samples	Range	Comment
Kingston, R.I., 1973-75[1]	6	1 - 15	Calculated as Aroclor 1254
Organ Pipe National Park, 1974[1]	6	0.02 - 0.41	No reference to Aroclor
Hays, Kansas, 1974[1]	3	0.03	No reference to Aroclor
Northwest Territories, Canada, 1974[1]	3	0.002 - 0.07	No reference to Aroclor
La Jolla, Calif., 1974[1]	6	0.5 - 14	No information on Aroclor
Vineyard Sound, Mass., 1973[1]	2	4 - 5	Calculated as Aroclor 1254
Univ. R.I., 1973[2]		2.1 - 5.8	Calculated as Aroclor 1254
Providence, R.I., 1973[2]		9.4	Calculated as Aroclor 1254
Chicago, Ill., 1975-76[3]	4	3.6 - 11.0	4% as 1242; 97% as "vapor"
Jacksonville, Fla., 1976[4]		3 - 36	
Lake Michigan, 1976-78[5]	6	0.57 - 1.6	74% as 1242; 88% as "vapor"
Milwaukee, Wis., 1978[5]	2	2.7	59% as 1254; 27% as 1260; 84% as "vapor"

SOURCE:
[1]Modified from Bidleman et al. (1978).
[2]Modified from Bidleman and Olney (1974).
[3]Modified from Murphy and Rzeszutko (1977).
[4]Modified from Fuller et al. (1977).
[5]Modified from Doskey (1978).

Rzeszutko calculated the amount of PCBs on particulates to be 27 percent of the total atmospheric burden.

Calculations based on another method for estimating vapor/particulate partitioning indicate that less than 5 percent of PCBs sampled in oceanic air could be associated with particulates (Junge 1977). Similar calculations for continental urban air indicate that 10 to 45 percent of Aroclor 1254 and 1260 and less than 5 percent of Aroclor 1242 could be associated with particulate matter. These estimates remain speculative because adsorption depends not only on the vapor pressure of individual mixtures, but also on the nature and amount of available surface area of the adsorbent, factors not known for most air masses.

Vapor-pressure data support the expectation that the more volatile isomers should preferentially accumulate in the atmosphere. PCB vapors with chlorine contents above a standard Aroclor 1016 mixture (greater than 41 percent chlorine) have been shown to contain more of the less chlorinated and thus more volatile isomers, for example the mono-, di- and trichlorobiphenyls (Mieure et al. 1975). This finding suggests that an atmospheric sample may bear only a slight resemblance to the composition of the original PCB product composition (Mieure et al. 1975).

Distribution of PCBs in Air

The current state of knowledge about the atmospheric cycle of PCBs is reflected in the summary of data on continental and marine air levels presented in Tables 1.5, 1.6, and 1.7. Relatively few measurements have been made of large-scale, geographically representative areas, comparing changes through time.

Atmospheric PCB concentrations for oceanic and rural continental areas range from 0.002 ng/m^3 to 1.6 ng/m^3, with more than 99 percent reported as existing in the vapor phase. Data obtained from atmospheric samples taken in the North Atlantic indicate an order of magnitude decrease in PCB levels between 1973 and 1977 (Bidleman, Department of Chemistry, University of South Carolina, personal communication, 1978). But measurements designed to investigate long-term trends in other locations do not exist, and thus it is impossible to ascertain whether these data reflect a global trend. A concentration of 0.05 ng/m^3 is used for calculating

TABLE 1.6 PCB Concentrations in Marine Air (10^{-9} g/m^3)

Location and Date	No. Samples	Range	Comments
Bermuda, 1973	4	0.15 - 0.5	Calculated as Aroclor 1254
Bermuda, 1973	7	0.19 - 0.66	Calculated as Aroclor 1254
Bermuda, 1974	25	0.08 - 0.48	Matched Aroclor 1248 most closely
Bermuda-U.S. Cruises 1972-74	17	<0.05 - 1.6	Matched Aroclor 1248 most closely
Chesapeake Bay, 1973	3	1.0 - 2.0	Calculated as Aroclor 1254
Grand Banks, 1973	5	0.05 - 0.16	Calculated as Aroclor 1254
Georges Bank, 1973	5	0.58 - 1.6	Calculated as Aroclor 1254

SOURCE: Modified from Bidleman et al. (1978).

TABLE 1.7 Trends in the Concentration of PCBs in N. Atlantic-Bermuda Air, Seasonal and Long-Term (Number of Samples)

Sampling Period	PCB - 1248 (10^{-9} g/m^3)
April 1973 - Sept. 1973	0.76 ± 0.55 (5), 4 were < 0.33
Oct. 1973 - March 1974	0.51 ± 0.53 (7), 5 were < 0.24
April 1974 - Sept. 1974	0.20 ± 0.09 (19)
April 1975 - Sept. 1975	0.076 ± 0.063 (6)
April 1976 - Sept. 1976	0.077 ± 0.035 (11)[a]
Oct. 1976 - March 1977	0.074 ± 0.099 (11)[a]

[a]These samples matched Monsantos Aroclor 1254 mixture more closely.

SOURCE: Compiled from Bidleman, Department of Chemistry, University of South Carolina, personal communication (1978).

the atmospheric PCB burden for oceanic and rural areas because the data converge around this value.

Measurements from urban/metropolitan areas also are quite variable, but data indicate concentrations ranging from 0.5 ng/m^3 to 36 ng/m^3. The average is about 4 ng/m^3 or 5 ng/m^3. Data on long-term trends do not currently exist for these areas.

Even less information exists about PCB levels in rain and snow, and it is difficult at present to predict levels for various geographically representative areas. The published data are presented in Table 1.4. Two data sets indicate that large variations can exist, both temporally and spatially. Whether the Lake Michigan measurements represent typical urban/metropolitan levels and whether the values measured at Norwegian sites represent those of remote regions is not certain.

Based on the above discussion the PCB burden in the atmosphere is conservatively estimated to be 18 x 10^{-3} kg using the following data:

Area of U.S. Atmosphere:	2,500 km x 6,500 km
Height of Assumed Affected Air Column:	2 km
Rural and Oceanic Area:	90 percent of affected area; 0.05 ng/m^3 concentration
Urban/Suburban Area:	10 percent of affected area; 5.0 ng/m^3 concentration

Removal of PCBs from the Atmosphere

PCBs are removed from the atmosphere via wet and dry deposition. The relative importance of each has not been resolved, mainly because of difficulties in measuring dry deposition on various surfaces. The most extensive work on wet deposition has been conducted in Lake Michigan (Murphy and Rzeszutko 1977). The applicability of these findings to other geographically representative regions has yet to be confirmed, but several important observations can be made at this time. In Lake Michigan, calculated yearly PCB deposition was found to be relatively uniform, although a large degree of variablility was observed in rain and snow samples. The weighted mean concentration of PCBs in rain was 120 ng/l, which indicates a gross flux to the lake of 0.83×10^{-4} g/m^2-yr (4,800 kg/yr).

The relationship between PCB concentrations in air and rain was studied using the concept of washout coefficients (Slinn et al. 1978). The washout coefficient for vapor depends upon the partitioning of PCBs between vapor and water, which is essentially Henry's Law Constant (H), whereas that for a particulate depends upon the collision efficiency of raindrop and particle, which is a function of the processes of diffusion, interception, and impaction. Because the precipitation-capture processes for the two phases are different, it is important to know the vapor/particulate partitioning (see discussion above).

The following relationship has been used to calculate vapor scavenging by rain: $W = \alpha p_0 \overline{C}$, where W = wet flux of PCBs, α = washout coefficient (concentration in rain divided by concentration in air, no density correction), p_0 = annual amount of rainfall, and $\overline{C}$ = average PCB concentration in air (Murphy and Rzeszutko 1977). Inspection of the Lake Michigan results revealed α to be 6.2×10^3 and 2.5×10^4 for Aroclors 1242 and 1254, respectively. A value of 7.8×10^4 as α, for total PCBs, has been obtained using data for dieldrin and DDT (Bidleman et al. 1978).

Washout coefficients also can be derived from $\alpha = RT/H$, where R = the universal gas constant (8.2×10^{-5} atm-m^3/mol-deg°K), T = absolute temperature, and H = Henry's Law Constant (Slinn et al. 1978). This would imply that H represents the effective Henry's Law Constant (because PCBs are mixtures) with values of 4 x

10^{-6} atm-m³/mol for Aroclor 1242 and 1 x 10^{-6} atm-m³/mol for Aroclor 1254.

The above results are two to three orders of magnitude lower than those obtained if the effective Henry's Law Constant for various Aroclors is calculated using both solubility and vapor-pressure data (Mackay and Leinonen 1975). In order to determine H, however, the vapor pressure and aqueous solubility must refer to material under similar physical conditions (Mackay, Department of Chemical Engineering, University of Toronto, Canada, personal communication, 1978). Most individual chlorinated biphenyl isomers are solid at room temperature whereas Aroclors, which are complex isomeric mixtures, are viscous fluids. These mixtures will be partitioned into individual components when present in air or water, and therefore, must be treated accordingly (Doskey 1978). Thus published Aroclor H values may be in error by two or three orders of magnitude (Mackay, Department of Chemical Engineering, University of Toronoto, Canada, personal communication, 1978). The calculated α coefficients from rain data (Murphy and Rzeszutko 1977) and volatilization rate experiments (Doskey 1978) seem to bear this out, but more extensive data are required. Nevertheless, it appears that the equation $W = \alpha p_0 \bar{C}$ gives a reasonable approximation for the wet deposition of PCBs (Murphy and Rzeszutko 1977).

Data on dry deposition fluxes, either to water or land surfaces, are not currently available. Measurements of PCB fallout have been made using glass plates covered with mineral oil and nylon or silicon screens coated with oil (Table 1.8). The data obtained using these techniques should not be considered representative of fallout rates, because the collection media do not reflect soil, vegetation, or water surfaces. However, such measurements can be useful in determining distributional patterns for various geographical locations.

HYDROSPHERIC COMPARTMENT

PCB contamination in the aquatic environment originates from several sources including precipitation, land runoff, industrial and municipal waste discharges, and accidental spills. Atmospheric transportation may be a significant factor in the global distribution of PCBs to

TABLE 1.8 Comparative Monthly Fallout of PCBs

Location and Date	Range (ng/m^2-month)	Comment
Sweden, 1969-71[1]	550 - 10500	As Clophen A50
Sweden, 1972-73[2]	320 - 2900	As Clophen A50
Iceland, 1971-73[3]	ND - 1050	Limit of Detection = 11
Czechoslovakia, 1972-73[6]	120 - 3400	As Clophen A50
Tsavo East, Kenya, 1972-73[6]	ND - 20	As Clophen A50
Nakuru, Kenya, 1972-73[6]	ND - 15	
Gambia[6]	ND	
Sri Lanka[6]	ND	
La Jolla, Calif., 1972[4]	9000	As Aroclor 1254
Southern Calif., Bizht, 1973-74[5]	3000	As Aroclor 1254

ND indicates not detected.

SOURCE:
[1]Modified from Sodergren (1973).
[2]Modified from Sodergren (1972).
[3]Modified from Bengtsson and Sodergren (1974).
[4]Modified from McClure and Lagrange (1977).
[5]Modified from Young et al. (1975).
[6]Modified from Sodergren (1975).

oceans and lakes, but coastal zones and estuaries near urban areas probably receive the major portion of PCB contamination from sources other than air (Young et al. 1977). PCBs initially deposited on land go through successive cycles of evaporation, deposition, and adsorption, until eventually these chemicals are carried via water runoff to lake and ocean sediments, the ultimate sinks.

When PCBs are introduced into the aquatic environment, they are dissolved in the water, adsorbed onto solids suspended in the water, or accumulated in different trophic levels (Sodergren 1973, Haque et al. 1974, Sayler and Colwell 1976, and Halter and Johnson 1977). PCBs have low water solubility and high octanol/water partition coefficients; thus, these compounds are adsorbed readily onto suspended solids, especially those high in organic carbon (Hamelink et al. 1971, Haque and Schmedding 1975, and Dexter and Pavlou 1978). Adsorption of PCB residues on the organic fraction of suspended solids and the subsequent incorporation of these particles into the sediment in aquatic systems is considered a major means for immobilizing PCBs (Oloffs et al. 1973, Haque et al. 1974, Moein 1976, Moein et al. 1976, and Hetling et al. 1978).

Mechanical and biological processes in aquatic systems may result in some recycling of PCBs. Degradation of material in the sediment may change its organic composition and release PCBs to the water column. Redistribution may also occur through erosion (Olsson et al. 1978). PCBs may be transported to the water column and surface film by the movement of sediment-generated bubbles, for example, hydrogen sulfide bubbles (Sodergren and Larssom 1979). It has been observed that a substantial portion (greater than 50 percent) of PCBs in pelagic carnivorous fishes apparently has been recycled through bottom sediment at least once (Weiniger 1978). Marine polychaetes are also able to accumulate PCBs from contaminated sediments (Courtney and Langstone 1978).

Despite major decreases in the influx of PCBs to coastal marine sites, only minor decreases in PCB concentrations can be observed in bottom sediment and flatfish (Young et al. 1977); this observation suggests continued recycling. During periods when contamination of the sediment becomes sufficiently high, PCBs may be released to the water column (Veith and Comstock 1975). If these releases were significantly greater than the rate of deposition, long-term pollution problems may result with the sediment acting as a source of PCB residues rather than a sink.

Data on deposition rates are scarce, and the basis for making estimates comes either from samples of layered sediment or from actual measurements of PCBs on settling particles. More studies are required because the available information is too scattered and incomplete to allow accurate determination of deposition rates.

Analysis of sandy, aerated marine sediment indicates that the less chlorinated isomers of a particular PCB mixture were not present, although this class of isomers could be found in anaerobic sediment samples (Eder 1976). Since it is unlikely that the less chlorinated forms would be retained differently in sediments, the findings suggest preferential decomposition. Observations in some Wisconsin rivers indicated that certain isomers of less chlorinated Aroclors disappeared downstream from the point source and that the proportion of highly chlorinated PCB compounds increased at the source (Veith 1972). This observation may also indicate preferential volatilization of the lighter Aroclor fractions.

PCBs can be degraded in activated sludge and it has been shown that certain bacteria can metabolize chlorobiphenyls (Ahmed and Focht 1973, Tucker et al. 1975, Tulp et al. 1978, and Furukawa et al. 1978). However, the highly chlorinated isomers are degraded more slowly than the less chlorinated forms. Thus, degradation of chlorobiphenyl isomers below penta- may occur in bottom sediment primarily as a result of microbial metabolism. Verification of this hypothesis is difficult, because field data are fragmentary and most laboratory studies have been conducted under conditions widely different from those of the natural aquatic environment. Available results do suggest, however, that degradation of most pentachlorobiphenyls is extremely slow and that of hexa- and more substituted chlorobiphenyls is practically negligible. The conclusion is that degradation is an insignificant factor in eliminating the highly chlorinated biphenyls from sediment.

PCBs in Fresh Water

The concentration of PCBs in fresh water (including suspended particles) is generally low. Even if the sediment contains appreciable quantities, the concentration in the overlying water may be below limits of detection using available analytical procedures. Median residue levels in the major drainage basins (major rivers) of the United States range from 0.1 μg/l to 3.0 μg/l (measured between 1971 and 1974: see Dennis 1976). Freshwater concentrations of PCBs in large lake systems have been observed to be much lower, probably due to their greater distances from point sources of PCBs. For the Great Lakes, the highest value reported in open waters has been for Lake Michigan: 31 ng/l (Table 1.9). The low values reported for Lake Superior and Lake Ontario were used as conservative estimates for freshwater levels of PCBs because the Great Lakes represent 98 percent of the total freshwater volume in the United States. The low and high estimates for

TABLE 1.9 PCBs in the Water of the Great Lakes

	ng/l
Lake Superior (as Aroclor 1254)	
1972-73	0.8[1]
1976	5.0[2]
Lake Michigan	31.0[3]
	9.0[4]
Lake Ontario	1.0-3.0[5]
Lake Huron	5.0-7.0[6]
Lake Erie	27.0[7]

SOURCE:
[1]Veith et al. (1977).
[2]Swain (in press).
[3]Haile (1977).
[4]Edie, NOAA-Great Lakes Environmental Research Laboratory, Ann Arbor, Michigan (unpublished data).
[5]de Lappe et al. (in press).
[6]Swain, Large Lakes Research Station, Grosse Ile, Michigan (unpublished data).
[7]Glooschenko et al. (1976).

freshwater levels in the four geographical areas used in this report are 1 ng/l and 3 ng/l (Table 1.10).

PCBs in Freshwater Sediment

Results obtained from monitoring sediment concentrations in European and African lakes and rivers are presented in Table 1.11. All PCB levels are reported on a dry weight basis. The lakes are divided into two groups: those subjected only to airborne contamination, and those contaminated with industrial and municipal wastes as well as fallout. The sediment concentration ranged from 4 μg/kg to 320 μg/kg for lakes receiving PCBs from all sources and 2 μg/kg to 20 μg/kg for those contaminated by atmospheric pollution only.

Sediment from the major drainage basins of the United States has been surveyed extensively for chemical contamination. From 1971 through 1974, median residue levels of PCBs in bottom sediment of 17 major rivers ranged as follows: 1.2 μg/kg to 64.5 μg/kg (1971), 2.0 μg/kg to 30.0 μg/kg (1972), 5.0 μg/kg to 50.0 μg/kg (1973), and 2.0 μg/kg to 160.0 μg/kg (1974). The

TABLE 1.10 Estimated PCBs in Fresh Water from Four Geographical Areas Across the U.S.

Area	H_2O Volume (km^3)	PCB Concentration (ng/l) low	high	Amount of PCB (kg) low	high
Pacific-Continental Divide	27	1	3	27	81
Appalachian Mountains-Atlantic Coast	100	1	3	100	300
St. Lawrence Drainage	11,320	1	3	11,320	33,960
Gulf of Mexico Drainage	129	1	3	129	387
			Total	11,556	34,728

highest residue levels were detected in the eastern portion of the United States (Table 1.12).

Bottom sediment from 26 streams discharging into the San Francisco Bay had been collected and analyzed (Law and Goerlitz 1974). PCB residues were found in all the streams surveyed with values ranging from 1 μg/kg to 1,400 μg/kg. The values found in two streams not located near industrial or commercial developments were greater than would be expected: 180 μg/kg and 610 μg/kg. No significant differences were found between the average residues of streams discharging into the northern or southern sections of the Bay.

From 1950 to 1977, the Hudson River in New York State received wastewater from two plants that used PCBs in producing transformers and capacitors. The plants appear to have been major contributors to contamination of the river; PCB concentrations in the sediment near the outfall from these plants ranged from 6.6 mg/kg to 67,000 mg/kg (Nadeau and Davies 1974, Hetling et al. 1978). The amount of PCBs stored in the sediment of the Hudson River is estimated to be approximately 300 x 10^3 kg (Toffelmire and Quinn, New York State Department of Conservation, personal communication, 1978).

As Figures 1.1 and 1.2 indicate, distribution of PCBs within the Hudson River sediment is not uniform. Maximum concentrations occur at depths of 5 cm to 30 cm rather than at the plane where water and sediment meet. Because the samples were collected from areas of high sediment accumulation, the observed depth distribution may reflect the reduction (beginning in 1976) and eventual cessation (in 1977) of PCB discharges from the

TABLE 1.11 PCB Residues in Sediment of Lakes and Rivers in Europe and Africa (No. Samples)

Atmospheric Sources		μg/kg dry weight
Lake Havgardssjon (10)	Sweden	10[3]
Lake Lyngsjo (3)	Sweden	8[3]
Lake Notsjon (3)	Sweden	20[3]
Lake Kinariochy (3)	Scotland	2[6]
All Sources		
Lake Vattern (20)	Sweden	170[1]
Lake Malaren (15)	Sweden	48[2]
River Viskan (10)	Sweden	45[2]
Lake Trummen (15)	Sweden	4[3]
Lake Oresjo (2)	Sweden	117[3]
River Kavlinge (5)	Sweden	30[3]
Lake Mcilwaine (10)	Rhodesia	120[5]
Hartbeespoort Dam (10)	South Africa	320[4]
Voelvlei Dam (10)	South Africa	70[4]
Lake Nakuru (10)	Kenya	20[7]

SOURCE:
[1]Modified from Hakansson (1977).
[2]Modified from Oden and Ekstedt (1976).
[3]Modified from Sodergren (unpublished).
[4]Modified from Greichus et al. (1977).
[5]Modified from Greichus et al. (1978b).
[6]Modified from Wells, Department of Agriculture and Fisheries of Scotland, personal communication.
[7]Modified from Greichus et al. (1978a).

General Electric plants at Hudson Falls and Fort Edward. If so, new, less contaminated sediment must have accumulated over older deposits. The values recorded are undoubtedly affected by erosion during and by the amount of backfilling that occurs after floods. Other factors in PCB deposition may be migration from sediment to overlying water columns and redistribution by microorganisms.

Deep penetration of PCBs has occurred where channels have been dredged and the rate of redeposition has been

TABLE 1.12 PCBs in Bottom Sediments in the Major Drainage Basins of the United States and Puerto Rico (Sample Size)

	1971		1972		1973		1974	
Drainage Basin	Median of Positive Detections in μg/kg	Range of Positive Detections in μg/kg	Median of Positive Detections in μg/kg	Range of Positive Detections in μg/kg	Median of Positive Detections in μg/kg	Range of Positive Detections in μg/kg	Median of Positive Detections in μg/kg	Range of Positive Detections in μg/kg
North Atlantic Slope	50 (3)	10 - 100	10 (56)	5 - 800	10 (58)	4 - 4000	31 (99)	2 - 800
South Atlantic Slope and Eastern Gulf of Mexico	64 (19)	10 - 200	30 (101)	5 - 500	10 (123)	5 - 600	11 (171)	2 - 530
Ohio River	9 (1)	9	-	-	20 (4)	20	35 (6)	9 - 61
St. Lawrence River	10 (1)	10	20 (10)	2 - 800	20 (24)	5 - 13000	20 (40)	3 - 700
Hudson Bay and Upper Mississippi River	-	-	ND (3)	ND	ND (14)	ND	ND (26)	ND
Missouri River	1 (2)	0.3 - 2	2 (22)	2	ND (17)	ND	ND (5)	ND
Lower Mississippi	30 (4)	10 - 80	2 (25)	2 - 2400	20 (57)	20	10.5 (81)	1 - 39

Western Gulf of Mexico	10 (20)	2 - 290	4 (66)	2 - 250	50 (188)	14 - 180	27 (157)	2 - 330
Colorado River	-	-	2 (7)	2	ND (4)	ND	ND (3)	ND
Great Basin	-	-	2 (9)	2	ND (7)	ND	ND (4)	ND
Pacific Slope Basins in California	-	-	20 (6)	2 - 190	ND (14)	ND	2 (26)	2 - 65
Pacific Slope Basins in Washington	-	-	2 (9)	2	ND (12)	ND	ND (5)	ND
Snake River	-	-	ND (1)	ND	ND (1)	ND	ND (4)	ND
Pacific Slope Basins in Oregon and Lower Columbia River	-	-	2 (1)	2	-	-	ND (1)	ND
Alaska	-	-	-	-	ND (2)	ND	-	-
Hawaii	-	-	ND (1)	ND	ND (1)	ND	-	-
Puerto Rico	-	-	-	-	5 (6)	5	160 (18)	10 - 640

- indicates no samples.
ND indicates not detected.

SOURCE: Modified from Dennis (1976).

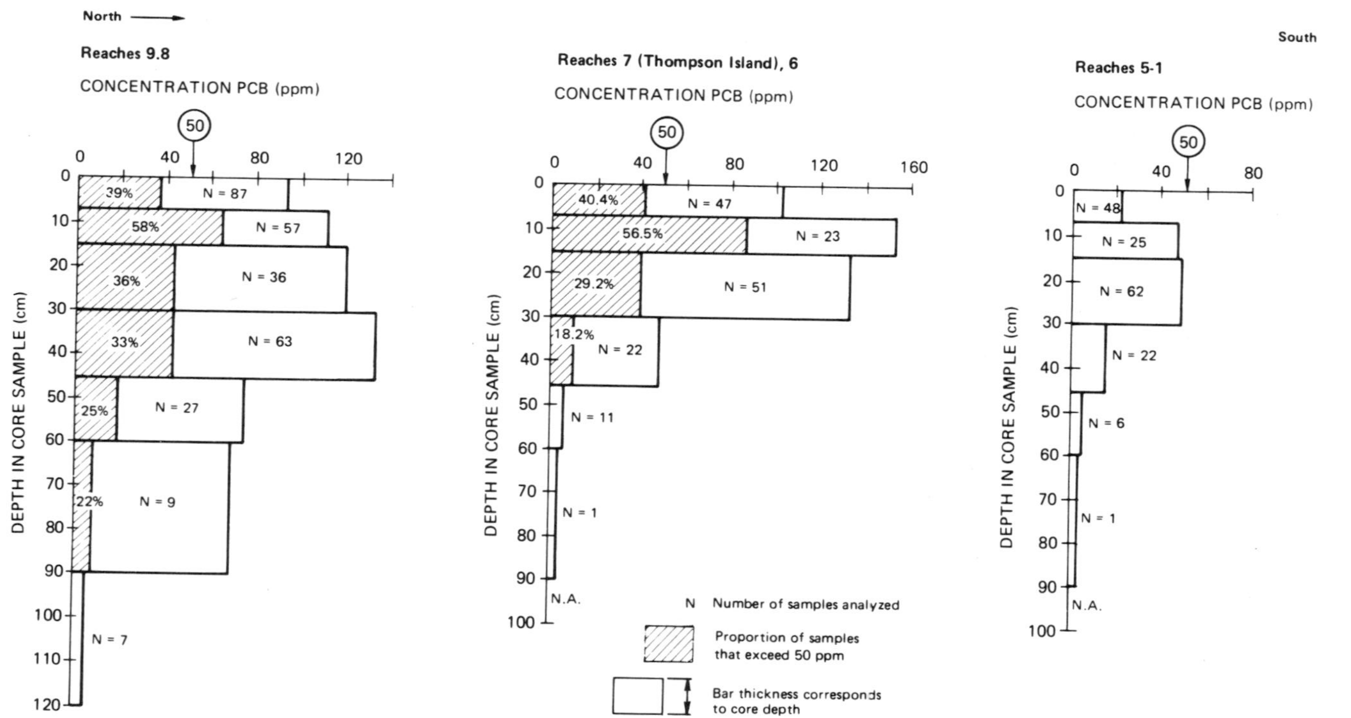

SOURCE: Sanders (1979).

FIGURE 1.1 Upper river detailed core network, 1975-1978. Summary of 672 PCB analyses of sediment samples (312 cores) from the upper Hudson River. Core samples were taken from river banks and shallows, not from deep-channel areas. Large pieces of woody debris were abundant in Reaches 9 through 6, but not common in Reaches 5 through 1.

as rapid as 30 cm/yr (Figure 1.2, Albany Basin, Germantown, and New York Harbor). Short core samples in coves and marginal shallow water areas from Kingston to Foundry Cove reflect sediment accumulation in areas not dredged.

During the period 1971 to 1972, the Chester River in the Upper Chesapeake Bay was monitored. PCB residues were found with sediment concentrations ranging from nondetectable to 300 μg/kg. The level detected at any one time depended on river flow rates (Clark and Murdock 1972). During 1974, bottom sediment was sampled from various lakes and rivers in Connecticut. The data indicated widespread contamination, particularly in highly developed areas. Levels up to 3,500 μg/kg were recorded (U.S. EPA 1976).

Table 1.13 summarizes the widespread contamination of freshwater bottom sediment throughout the United States (Crump-Wiesner et al. 1973). Samples were taken from lakes and streams remote from industrial centers to represent regional conditions. Twenty percent of the sediment samples taken across the nation contained PCBs with levels ranging from 5 μg/kg to 2,400 μg/kg.

The occurrence and distribution of PCBs in sediment from the Great Lakes have been studied extensively (Veith 1975, Glooschenko et al. 1976, Haile 1977, Veith et al. 1977, PLUARG 1978, Swain 1979, Eisenreich et al. 1979). Table 1.14 illustrates the concentrations detected in specific lake systems. These data seem consistent with sediment data from other sections of the country.

An evaluation of the degree of PCB contamination throughout the United States indicates regional variation. In Table 1.15 residue data from the United States Geological Survey (USGS) are summarized for 1972 through 1974. These data show an increase in the geographical dispersion of PCBs. In 1972, 5 states measured PCBs in bottom sediment; in 1973, 8 states, and in 1974, 13 states reported contamination with these compounds (U.S. EPA 1976). Those states reporting PCB contamination during this period continue to detect levels of the chemical today. Data from U.S. monitoring programs and from reports in the literature through 1977 indicate that the highest level of PCBs in freshwater sediment is found in heavily industrialized, populous areas, particularly in the eastern part of the country. The lowest values are found in rural areas.

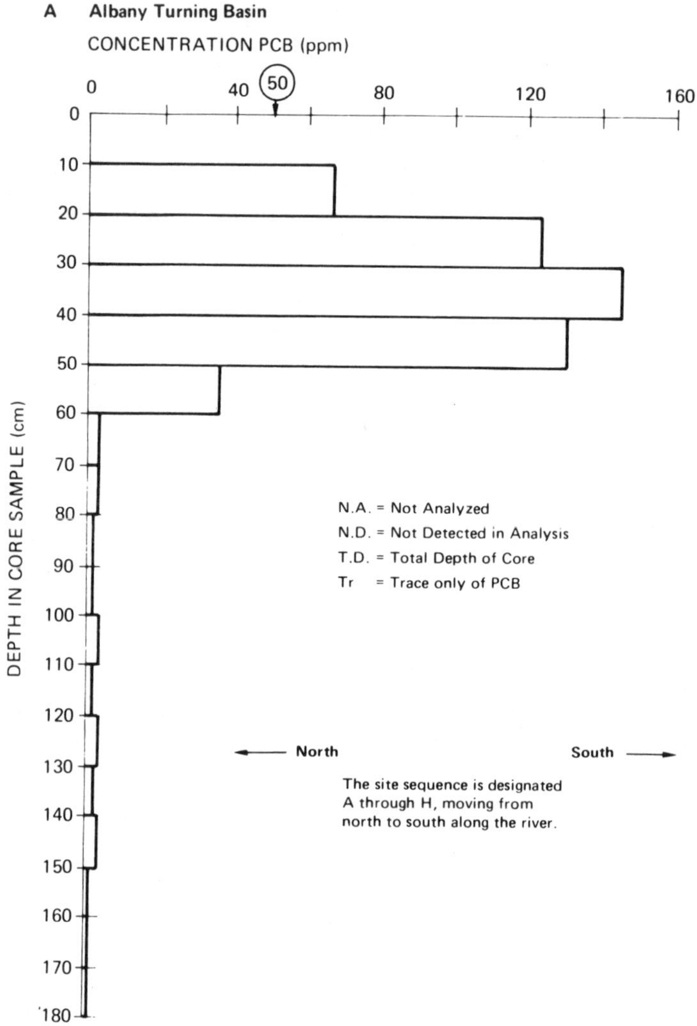

SOURCE: Sanders (1979).

FIGURE 1.2 Single core reconnaissance of the Hudson River. Summary of PCB analyses from eight single-core samples from a reconnaissance traverse of the lower river regions from Albany to New York Harbor. Each bar represents two analyses, one for each of two different PCB isomers (Aroclor 1242 and 1254).

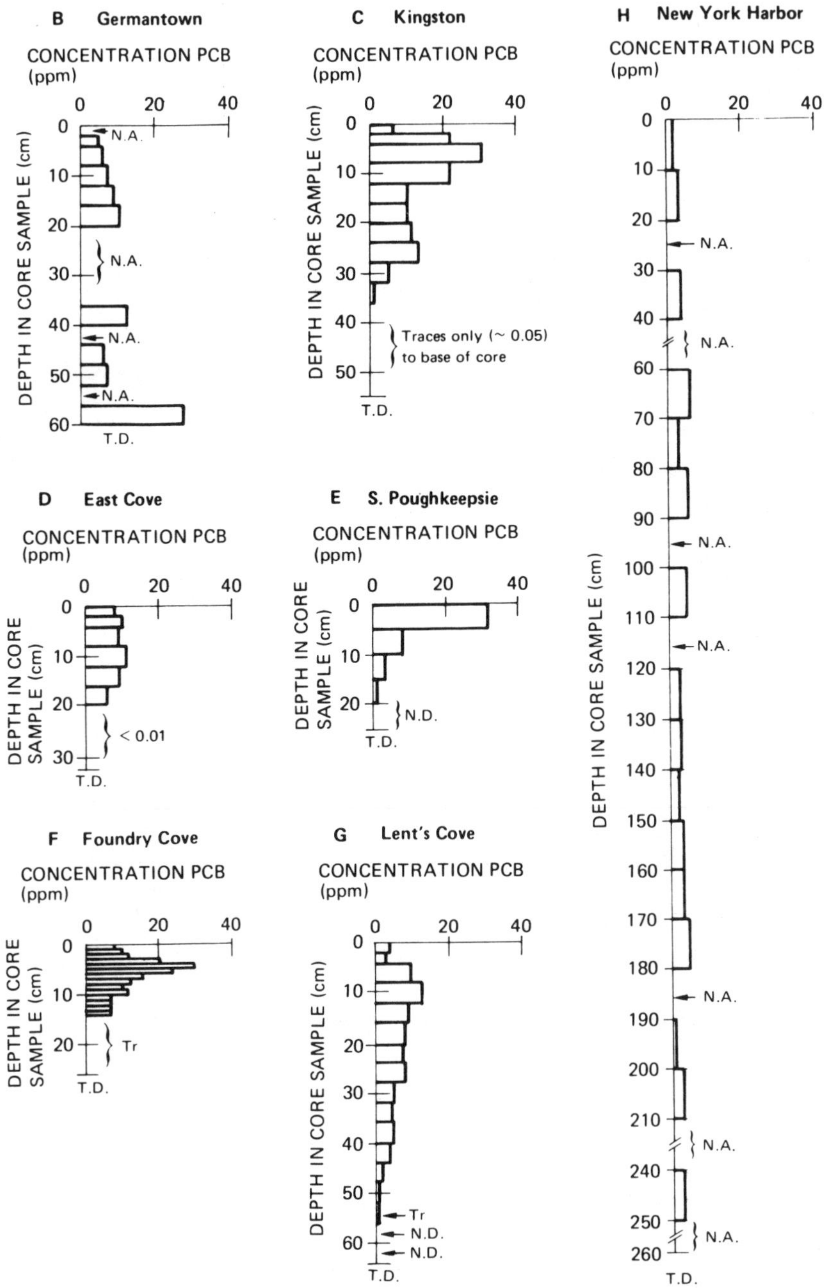
B Germantown
CONCENTRATION PCB (ppm)
0 20 40
DEPTH IN CORE SAMPLE (cm)
0 10 20 30 40 50 60
N.A.
N.A.
N.A.
N.A.
T.D.
C Kingston
CONCENTRATION PCB (ppm)
0 20 40
DEPTH IN CORE SAMPLE (cm)
0 10 20 30 40 50
Traces only (~ 0.05) to base of core
T.D.
D East Cove
CONCENTRATION PCB (ppm)
0 20 40
DEPTH IN CORE SAMPLE (cm)
0 10 20 30
< 0.01
T.D.
E S. Poughkeepsie
CONCENTRATION PCB (ppm)
0 20 40
DEPTH IN CORE SAMPLE (cm)
0 10 20
N.D.
T.D.
F Foundry Cove
CONCENTRATION PCB (ppm)
0 20 40
DEPTH IN CORE SAMPLE (cm)
0 10 20
Tr
T.D.
G Lent's Cove
CONCENTRATION PCB (ppm)
0 20 40
DEPTH IN CORE SAMPLE (cm)
0 10 20 30 40 50 60
Tr
N.D.
N.D.
T.D.
H New York Harbor
CONCENTRATION PCB (ppm)
0 20 40
DEPTH IN CORE SAMPLE (cm)
0 10 20 30 40 60 70 80 90 100 110 120 130 140 150 160 170 180 190 200 210 240 250 260
N.A.
N.A.
N.A.
N.A.
N.A.
N.A.
N.A.
T.D.

TABLE 1.13 Summary of PCB Residue Data for Bottom Sediments January 1971 - June 1972 (Number of Samples)

State	Range (μg/kg)	Median Concentration (μg/kg)
Alaska (3)	---	---
Arkansas (23)	20 - 2,400	60
California (13)	20 - 190	85
Connecticut (1)	40	---
Hawaii (4)	---	---
Georgia (12)	10 - 1,300	300
Maryland (11)	10 - 1,200	30
Mississippi (8)	50; 170	---
New Jersey (12)	8 - 250	20
Oregon (4)	15; 140	---
Pennsylvania (16)	10 - 50	20
South Carolina (11)	30 - 200	50
Texas (293)	7.9 - 290	80
Virginia (10)	5 - 80	40
Washington (10)	---	---
West Virginia (2)	10	---

SOURCE: Modified from Crump-Wiesner et al. (1973).

Table 1.16 presents estimates of total PCB burdens in sediment for particular regions throughout the United States: the Pacific Coast to the Continental Divide, the Appalachian Mountain Range to the Atlantic Coast, the St. Lawrence drainage system including the north-central and Great Lakes region, and the Gulf of Mexico drainage system including the south-central region of the United States. When calculating a total PCB burden for sediment, it is necessary to consider possible mixing activity of bottom-living organisms. This activity may carry PCB residues deeper into the sediment than would be otherwise anticipated. For that reason, a depth of 15 cm has been used for potential PCB

TABLE 1.14 PCBs in Sediment of the Great Lakes (No. Samples)

		µg/kg
Lake Superior		30.0[1]
	(4)	7.0±0.5[2]
	(17)	tr - 20.0[3]
Lake Michigan		38.0[1]
	(7)	12.0-25.0[4]
Lake Huron		9.0-33.0[1]
	(9)	tr - 20.0[3]
Lake Erie		74.0-250.0[1]
Lake Ontario		77.0-89.0[1]
	(7)	43.0-240.0[4]

tr indicates trace.

SOURCE:
[1]PLUARG (1978).
[2]Veith et al. (1977).
[3]Glooschenko et al. (1976).
[4]Haile (1977).

contamination. With an average sedimentation rate of 3.3 mm/yr (a conservative estimate based on values obtained in the Great Lakes), 15 cm represents the average amount of sediment accumulated during the last 50 years (Edgington and Robbins 1976).

The area from the Pacific Coast to the Continental Divide seems to have the lowest PCB level in sediment. Most recorded values are below 10 µg/kg; therefore 2 µg/kg to 20 µg/kg are chosen as low and high estimates of PCB levels in sediment from this region. Thus, the calculated PCB burden in sediment for the region from the Pacific Coast to the Continental Divide is from 5 x 10^3 kg to 65 x 10^3 kg (Table 1.16).

The Appalachian Mountain-Atlantic Coast region and the St. Lawrence drainage basin represent areas where consistently high levels of PCBs have been recorded in various components of the aquatic ecosystem. The Hudson River is the most contaminated water system in the Atlantic Coast area: approximately 300 x 10^3 kg of PCBs have been deposited in the river sediment (Toffelmire and Quinn, New York State Department of Conservation, personal communication, 1978). Low and high estimates of PCBs in freshwater sediment for the Appalachian-

TABLE 1.15 Summary of PCB Residue Data for Bottom Sediments (Means and Ranges in µg/kg)

	1972		1973		1974	
	Mean	No. of Stations	Mean	No. of Stations	Mean	No. of Stations
Arkansas	0	6	0	6	12.8 (<2-29)	15
California	0	1	0	15	33.5 (<2-65)	18
Connecticut	127.5 (<5-E800)	16	14.1 (<5-40)	16	71.5 (<2-350)	40
Florida	93.6 (<5-400)	27	35.8 (<2-600)	64	43.3 (<2-530)	68
Georgia	---	---	0	1	13.6 (<1-51)	9
Illinois	---	---	20	1	---	---
Indiana	---	---	---	---	35	4
Michigan	2	1	---	---	17.6 (<4-60)	13

Mississippi	---	---	0	1	16.4 (<3-56)	50
New Jersey	--- (<5,<10)	2	230.2* (<3-4000)	20	172.2 (<1-800)	15
New York	---	---	519* (<2-1300)	17	95.8 (<2-450)	17
Pennsylvania	50.7 (<5-E100)	9	11.7 (<5-20)	5	105.6 (<6-700)	23
Texas	61.7 (<4-120)	41	17.6 (<3-45)	58	45.5 (<1-153)	58
Wisconsin	---	---	10	9	0	21
Puerto Rico	---	---	0	6	242 (<10-640)	16

--- indicates no samples.
E indicates estimated.
* indicates mean figure does not include anomalous values.

SOURCE: Modified from Nisbet (1976).

TABLE 1.16 Estimated PCB Levels in Sediment from Four Geographical Areas Across the U.S.

Area	Sediment Volume (km^3)	PCB Concentration (μg/kg) low	high	Amount of PCBs (x 10^3 kg) low	high
Pacific Coast-Cont. Div.	3.2	2	20	5	65
Appal. Mount.-Atlantic	5.8	100	500	580	2900
St. Lawrence Drainage	23.7	30	150	700	3500
Gulf of Mexico Drainage	10.5	10	50	100	500

NOTE: The following assumptions are made - the sediment is mainly organogenic with a water content of approximately 90 percent. The PCB residues are stored within the upper 15 cm of the sediment layer.

Atlantic Coast area are 100 μg/kg and 500 μg/kg, respectively with the associated PCB burden ranging from 580 x 10^3 kg to 2,900 x 10^3 kg (Table 1.16).

The Great Lakes are not only the dominant water areas of the St. Lawrence drainage system, but they also constitute approximately 98 percent of the total freshwater volume of the United States. Based on the results presented in Table 1.14, 30 μg/kg to 150 μg/kg are chosen as low and high estimates for PCBs in sediment. The estimated range of PCB burden in the sediment of the St. Lawrence drainage area amounts to approximately 700 x 10^3 kg to 3,500 x 10^3 kg, thus representing the greatest PCB reservoir in the continental United States.

The drainage area of the Gulf of Mexico contains sediment with concentrations that range from non-detectable to 530 μg/kg, excluding samples taken in the vicinity of heavy industrial contamination. Samples below 30 μg/kg were in the majority; thus 10 μg/kg and 50 μg/kg are considered reasonable low and high estimates, indicating a PCB load of 100 x 10^3 kg to 500 x 10^3 kg stored in the bottom sediments.

PCBs in Freshwater Biota

Accumulation of PCBs in freshwater fish has been of special concern because fish are a major source of PCB contamination in the diet of humans. Concentrations in certain fish species high in fat content have been in excess of the FDA tolerance level of 5 mg/kg.

The occurrence of PCBs in fish differs among species, as would be expected, due to differences in food gathering habits, behavior patterns, and in content of body fat. Even when fish samples are taken from the same environment, PCB concentrations in species with large amounts of fat tissue, such as carp, catfish, lake trout, or coho and chinook salmon, are greater than those in low-fat species, such as yellow perch, walleye, or northern pike. In a 1975 study using fish samples from Green Bay, Lake Michigan, PCB concentrations in the edible tissues ranged from 10.0 mg/kg to 20.0 mg/kg in lake trout, 5.1 mg/kg to 51.6 mg/kg in carp, and 3.2 mg/kg to 5.6 mg/kg in yellow perch (Kleinert 1976). The lake trout and carp had a fat content in excess of 10 percent of body weight, whereas the yellow perch fat content was less than 4 percent.

Large differences in concentrations of PCBs have been reported in fish taken from the Hudson River. For example, yellow perch had PCB levels of 16.0 mg/kg to 38.0 mg/kg and the content in goldfish ranged from 45 mg/kg to 900 mg/kg (New York State Department of Environmental Conservation 1978). Elsewhere in various fresh waters of New York State, the PCB content of fish ranged generally from 0.2 mg/kg to 2.1 mg/kg, although a few species had as much as 14 mg/kg.

PCB concentrations within species can vary geographically. For example, coho salmon samples taken from the Wisconsin portion of Lake Michigan had PCB levels of 0.02 mg/kg to 26.0 mg/kg (Kleinert 1976); those from the New York waters of Lake Erie had 0.9 mg/kg to 5.7 mg/kg of PCBs; 1.7 mg/kg to 7.3 mg/kg of PCBs were detected in samples taken from New York waters of Lake Ontario (New York State Department of Environmental Conservation 1978).

Data from various sources indicate that the PCB concentrations for species in certain environments, such as Lake Michigan, the upper region of the Mississippi River, the Fox River in Wisconsin, and the Hudson River in New York State, exceeded the FDA tolerance level of 5 mg/kg. These waters are known to have received large amounts of PCBs from industrial wastes in the past. The PCB content in fish in the upper Mississippi River has declined, but no such trend is evident in fish from the Great Lakes (Kleinert 1976).

Based on a maximum standing fish crop in the Great Lakes of 5×10^9 kg and a mean PCB level of 3 mg/kg, the PCB content of freshwater fish in the United States is estimated to be 15×10^3 kg (Veith 1975, Graham 1976, and Veith et al. 1977). It seems reasonable that the remainder of the Great Lakes biota would carry an amount approximately equal to this, giving a high estimate of 30×10^3 kg. Other U.S. freshwater biota and associated PCB levels are negligible in comparison.

PCBs in the Oceans

Although some data exist on samples taken from the Pacific Ocean and the Gulf of Mexico, most measurements of PCBs in marine waters have used samples collected in the North Atlantic Ocean. A summary of most of the published data is presented in Table 1.17.

Clear-cut explanations of the observed spatial and temporal trends remains elusive. It is especially difficult to explain measurements which suggest that PCB concentrations in North Atlantic surface waters dropped by a factor of more than 40 between the summers of 1972 and 1973 (Harvey et al. 1974). The data have generated lively debates, but a satisfactory explanation of them has yet to emerge (Duce and Duursma 1977). Consequently, it is fair to say that we have a poor understanding of historical trends in marine levels of PCBs. This limitation is unfortunate because it prevents accurate predictions of residence times and ultimate fates of PCBs in an important environmental pool.

A review of the data presented in Table 1.17, however, does indicate generally higher levels of PCBs in the North Atlantic than in the Pacific Ocean and the Gulf of Mexico. Airborne PCBs may enter the North Atlantic Ocean mainly between latitudes 40° and 45°N as a consequence of the southwesterly winds prevailing from the United States (Harvey and Steinhauer 1976b). Speculations have been made on the relative importance of the several ways to distribute PCBs within the oceans (Harvey and Steinhauer 1976a). These include: (1) dilutions of PCBs by vertical and horizontal advection and diffusion; (2) association of PCBs with descending particulate matter; (3) evaporative codistillation of PCBs from surface waters in low- and mid-latitudes; and (4) biogeochemical degradation of PCBs in the mixed layer. Currently, however, no quantitative information exists on the relative importance of each process. Particle removal is probably the most important pathway for moving PCBs from the mixed layer to bottom sediment, but few measurements of PCB accumulation rates have been published (Harvey and Steinhauer 1976a).

For purposes of calculating the oceanic PCB load, an average concentration of 2.6 ng/l was used for estimating the amount of PCBs in the upper 200 m of the North Atlantic Ocean (see Table 1.17: surface values are 2.6 ng/l and mixed layer values are reported as 3.6 ng/l and 1.8 ng/l). A conservative estimate of 1 ng/l was used as the amount of PCBs in water depths below 200 m (Table 1.17 indicates reported values of 2 ng/l and 1.3 ng/l). The area of the North American Basin was taken as 10×10^{12} m^2 with a depth of 4,200 m (Harvey and Steinhauer 1976a). Calculations using these data suggest that North American Basin waters contain an upper limit of 6.6×10^7 kg of PCBs. Calculations by other investigators suggest a burden of 4.2×10^7 kg (Harvey and Steinhauer 1976a). However, it has been suggested that all measurements of PCBs in oceans have been biased upwards by sampling contamination, and that reported measurements may be too high by at least one order of magnitude (Risebrough et al. 1976). Accordingly, we have adopted as a lower limit of PCB content the value of 6×10^6 kg.

PCBs in Marine Sediment

PCBs are transported to ocean waters adsorbed on suspended organic particles in river water. Sewage sludge dumped along coastal areas also contributes to the concentration as does deposition of polluted dredged spoils. It is estimated that approximately 4×10^9 kg of sewage sludge and 13×10^9 kg of dredged spoils are dumped off the Atlantic and Gulf Coast annually (Council on Environmental Quality 1970). If an average PCB concentration of 400 μg/kg wet weight is assumed (Berg and Peoples 1977), approximately 7×10^3 kg of PCB residues are transferred to the coastal waters. Annual

TABLE 1.17 PCB Concentration of Oceanic Water (ng/l)

Date	Location and Type	Range	Mean	Reference
3/73	Saragasso Sea (surface)	<.9 - 3.6	<1.5	Bidleman and Olney (1974)
9/73	Saragasso Sea - N.Y. (surface)		.8	Harvey et al. (1974)
9/73	Azores-Barbados (surface)		2	Harvey et al. (1974)
1973/74	Gulf of Mexico (surface)		1.8	Giam et al. (1976)
1974	British Coastal waters (surface)	<.15 - .52	0.23	Dawson and Riley (1977)
9/75	Iceland-New England (surface)	1-7	4	Harvey and Steinhauer (1976a&b)
9/75	South Atlantic (11° S - 35° S) (surface)	0.3 - 3.7	0.9	Harvey and Steinhauer (1976a&b)
1972	North Central Pacific Gyre (surface)		5	Williams and Robertson (1975)
1973	Calif. current (surface)	0.3 - .5		Young (1975)
1974	New England Continental Shelf (surface)		0.8	Harvey et al. (1974)

1975	Atlantic Northeast Trades (surface)	3.5 - 5		Harvey and Steinhauer (1976a&b)
1973-75	North Atlantic (surface)	0.3 - 8.0	2.6	Harvey and Steinhauer (1976a&b)
1975	Pacific offshore waters (surface)	ND[a]		Risebrough et al. (1976)
9/73	Western boundary of Southern Calif. bight (surface)	0.27 - 0.49	0.4	Risebrough et al. (1976)
5/73	Calif. Coastal waters (surface)	1.1 - 5.9	2.1	Risebrough et al. (1976)
2/75	Atlantic Northeast Trades (mixed layer)[b]	0.9 - 8.7	3.6	Harvey and Steinhauer (1976a&b)
9/75	North Atlantic (mixed layer)	1.7 - 1.8	1.8	Harvey and Steinhauer (1976a&b)
2/75	Atlantic Northeast Trades (deep water)[c]	0.8 - 4.2	2.0	Harvey and Steinhauer (1976a&b)
9/75	North Atlantic (deep water)	0.7 - 2.4	1.3	Harvey and Steinhauer (1976a&b)

[a]Limit of detection ranged from 0.03 - 0.3 ng/l.
[b]Mixed layer is taken as the upper 200 meters.
[c]Deep water is taken as below a depth of 200 meters.

PCB transport from U.S. rivers to the oceans is estimated at 2 x 10^5 kg (Nisbet and Sarofim 1972).

Data on PCB levels in marine sediment are available primarily for coastal areas, estuaries, and harbors. Measurements in sediment deposited on the continental shelves seem to be lacking. Whereas most of the PCBs delivered to the open ocean remain in solution or are adsorbed onto suspended solids, it is likely that most of the PCBs delivered to the coastal regions are deposited in the sediment. The greater abundance of suspended organic solids in coastal waters (100 times the amount in the open ocean) plus a more rapid rate of deposition contribute to the removal of a larger proportion of PCBs from the water column in these areas. Therefore, it is inferred but not yet proven that most of the PCBs deposited in marine sediment will be found within the continental terraces.

<u>The Gulf of Mexico</u> Because the Gulf of Mexico receives sediment runoff from about two-thirds of the U.S. land area and one-half the area of Mexico, the content of PCBs in sediment of the Gulf is representative of the runoff situation for a large area of the continent (Table 1.18). PCB levels from 0.2 μg/kg to 35 μg/kg were found in sediment sampled in coastal areas of Texas and Louisiana (U.S. EPA 1976). In Escambia Bay, Florida, where sediment has been sampled since 1969, it has been established that an industrial outfall is responsible for part of the PCB contamination in that area. Sediment sampled near the outfall reached a PCB level of 486 mg/kg in 1969, but the contamination appears to have decreased since then (Table 1.19 and Figure 1.3).

<u>The North Atlantic Ocean</u> In a preliminary survey in 1973 in St. Margaret's Bay, Nova Scotia, investigators were unable to detect any PCB residues (limit of detection 1 μg/kg) in sediment samples taken from water depths of 5 m to 20 m and 60 m to 75 m (Hargrave and Phillips 1975). In surface sediment off Long Island and New Jersey, PCB levels were found at 10 μg/kg and 40 μg/kg, respectively (West and Hatcher, Atlantic Oceanographic Research Laboratory, NOAA, personal communication, 1978). In a dump site in the Christiensen Basin, at the head of Hudson Shelf Valley (73°45'W; 40°27'N, ESE of harbor light entrance to the New York Harbor), one core sample revealed 2,000 μg/kg

TABLE 1.18 Concentrations of PCBs in Selected Samples from the Gulf of Mexico (1971-1975)

	Year of Sampling	Number of Stations	ng/g
Biota	1971-72	44	56.0
	1973-74	24	26.0
Sediment	1973-75	36	0.3
Water	1973-74	34	1.8
Air	1973-75	8	0.4

SOURCE: Taken from Giam et al. (1976).

TABLE 1.19 Residues of Aroclor 1254 in Sediment Samples from Escambia Bay and River (mg/kg) (See Figure 1.1 for Station Locations)

Location	Depth	1970	1971	1972
1				
2				
6				
8		ND		
12		1.9		
13		1.8		
14		30.0		
15		4.2		
16		3.3		
17		4.9		
18		0.6		
19		2.5		
20		1.4		
21	0 - 5 cm	78.0	8.1	0.97
	5 - 10 cm	30.0	0.12	5.8
	10 - 15 cm	6.1	ND	---
	15 - 20 cm	0.4	ND	---
22	0 - 5 cm	10.0	0.91	0.14
	5 - 10 cm	11.0	ND	ND
	10 - 15 cm	15.0	ND	ND
	15 - 20 cm	20.0	ND	---
	20 - 25 cm	18.0	ND	---
	25 - 30 cm	1.2	ND	---
23	0 - 5 cm	0.19	0.19	---
	5 - 10 cm	0.08	ND	---
	10 - 15 cm	0.02	ND	---
	15 - 20 cm	ND	ND	---

ND indicates not detected.

SOURCE: Modified from U.S. EPA (1976).

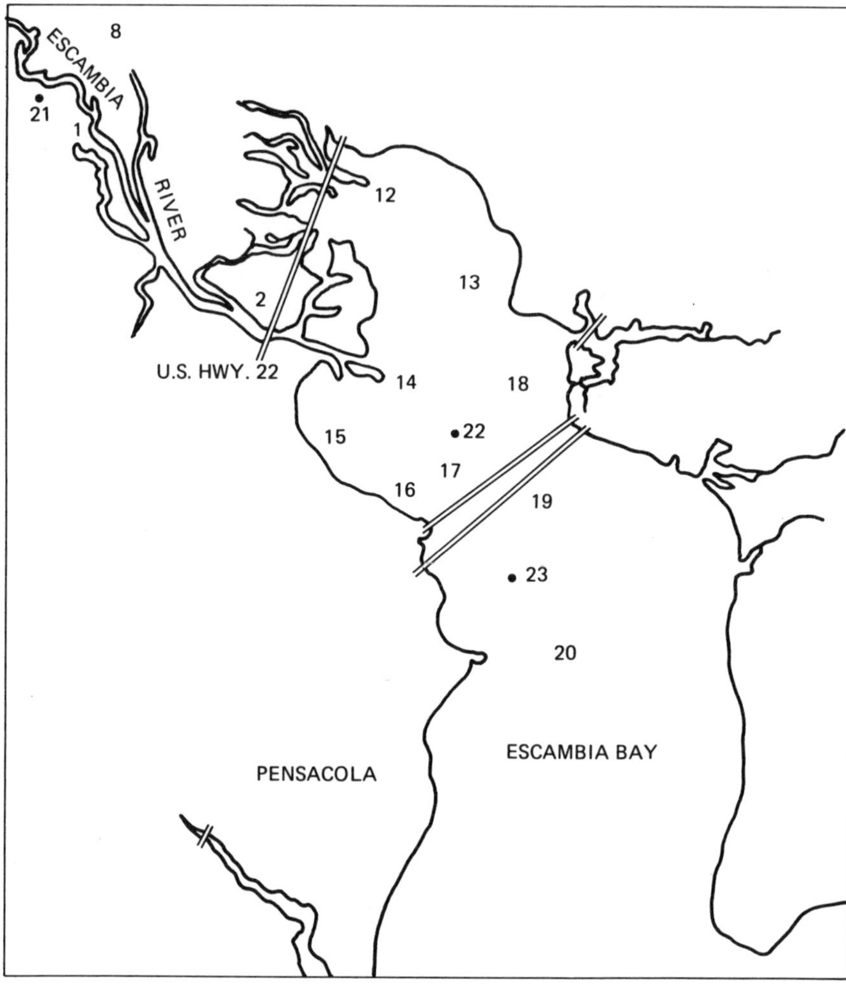

SOURCE: Modified from U.S. EPA (1976).

FIGURE 1.3 Escambia Bay sampling stations.

in the upper 4 cm and 400 μg/kg in a zone 4 cm to 23 cm below the sediment surface. In coastal sediments off Nova Scotia, New Brunswick and in the Northumberland Strait, PCBs were near or below the detection limit of 5 μg/kg (Leonard 1977). However, in a 1976 survey of four major harbors in the Atlantic Provinces of Canada, levels of PCBs ranged from 10 μg/kg to 2,900 μg/kg

(Travers and Wilson 1977). In deep sea sediments off northern Canada, no PCBs were found (detection limit of 5 μg/kg) for Baffin Bay at water depths of 750 m to 2,012 m, Lancaster Sound at depths of 520 m to 570 m, and Smith Sound at 534 m (Pocklington, Bedford Institute, Halifax, Nova Scotia, personal communication, 1978).

Pacific Ocean The coastal waters of southern California have been extensively sampled for evidence of PCB residues. Dated sediment from the Santa Barbara Basin indicate that deposition of PCBs began about 1945 (Hom et al. 1974). Through 1967, the levels of PCBs detected were increasing continuously (Table 1.20). Under anaerobic conditions burrowing organism activity is negligible so that PCBs can remain undisturbed in sediment for at least 30 years.

Despite major decreases in PCB emissions over a 3-year period by a dominant discharger to a coastal marine site in Los Angeles, California, only minor decreases in PCB levels were observed in the bottom sediment (Young et al. 1977). In areas near the outfall, levels varied from 7 mg/kg in the top 5 cm in 1972 to about 4 mg/kg in 1975. In areas 12 km from the outfall, the corresponding values were 0.5 mg/kg in both 1972 and 1975.

Sediment samples taken off the Peruvian Coast at depths of 510 m and 915 m did not have detectable PCB residues; the limit of detection was 5 μg/kg (Pocklington, Bedford Institute, Halifax, Nova Scotia, personal communication, 1978).

That PCBs can be transported to considerable water depths is indicated by their presence in the

TABLE 1.20 PCB Levels in Dated Sediments from the Santa Barbara Basin (μg/kg, dry weight)

1940 - 1945	31
1947 - 1952	49
1955 - 1960	66
1962 - 1967	100

SOURCE: Reprinted (with modifications) with permission from [Science 184:1197-1199, Hom, W., R.W. Risebrough, A. Soutar, and D.R. Young, Deposition of DDE and PCB in dated sediments of the Santa Barbara Basin], Copyright [1974], American Association for the Advancement of Science.

bathydemersal fish, Antimore rostrata, a permanent resident of the lower continental slope in Pacific waters (Risebrough et al. 1976). These fish swim slowly over the bottom, crop benthic epifauna, and root in the sediment for infauna. The PCB residues found in tissue samples taken from these fish are believed to originate from the sediment.

The North Sea Sediment from the central North Sea and the Norwegian Depression has been sampled (Eder 1976) at water depths of about 50 m and 450 m, respectively. In the upper 20 cm of the sediment, PCB levels averaged 28 μg/kg in the North Sea; in the anaerobic sediment of the Norwegian Depression,levels of 56 μg/kg were noted (Table 1.21). The sampling sites in the North Sea are sufficiently remote from any point sources of direct pollution suggesting that atmospheric transport of PCB residues must have been the source. It also was observed that the percentage of less chlorinated PCBs was higher in the anaerobic sediment of the Norwegian Depression than in the well-aerated sand sites in the North Sea. If the ratios of high to low chlorine isomers were identical for both areas at the time of deposition, this observation would suggest that biological degradation may have occurred in the sand

TABLE 1.21 PCB Concentrations in the Central North Sea and Norwegian Depression

Site Number	I	II	III	IV
Latitude	55°30'N	55°30'N	57°48'N	57°49'N
Longitude	06°30'N	04°50'E	07°10'E	08°29'E
Depth below sea level	50 m	45 m	420 m	505 m
Sediment Type	Homogeneous silty fine sand	Fine to medium sand	Clay and silt	Viscous clay and silt
PCB (Clophen) (weighted average, μg/kg dry sediment)				
Upper 10 cm	20.4±3.6	14.0±2.8	28.0±33.9	23.7±6.3
Below 10 cm	11.5±2.8	10.5±12.3	40.5±30.9	19.3±6.7

SOURCE: Reprinted (with modifications) with permission from [Chemosphere 2:101-106, Eder, G., PCBs and compounds of the DDT group in sediments of the central North Sea and the Norwegian Depression], Copyright [1976], Pergamon Press, Ltd.

biotope. Physical factors related to solubility and diffusion may be important also and could result in PCB transport from sediment.

<u>The Baltic Sea</u> From 1972 through 1973, sediment taken from various areas and depths in the Baltic Sea was sampled and analyzed for PCB residues (Oden and Ekstedt 1976). Only surface sediments were studied (0 to 2 cm). In the northern part of the Baltic Sea, the PCB content was fairly low (about 20 μg/kg) and without large geographical variation (Figure 1.4). Similarly low

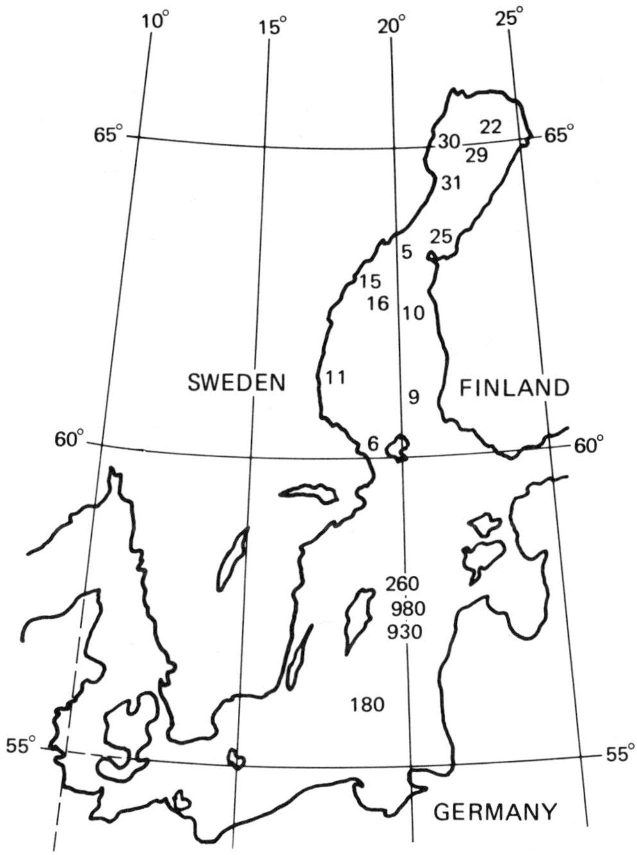

SOURCE: Modified from Oden and Ekstedt (1976).

FIGURE 1.4 PCB concentrations in sediment samples taken in the Baltic Sea (μg/kg, dry weight).

levels had been observed in earlier studies (Linko et al. 1974). The high contamination (about 980 μg/kg) found in the central and southern portions was most probably the result of waste disposal practices.

The marine sediment data presented in the above subsections were used to estimate the quantity of PCBs stored in coastal areas of the United States. We have inferred that most PCBs are accumulated in sediment within 30 km from the coast (within 2 km on the Pacific Coast). The PCB burdens indicated in Table 1.22 were calculated for the upper 5 cm of the sediment layer.

PCBs in Marine Biota

The mass of marine biota over the eastern continental shelf of North America is taken to be 10 percent of the global marine biota, or about 1.5 x 10^{11} kg (National Research Council 1977). Using a concentration of 0.2 mg/kg (Risebrough et al. 1972, Graham 1976), a plausible upperbound estimate of PCBs in North Atlantic marine biota is 3 x 10^{4} kg.

Removal from the Hydrosphere

In the course of determining the mass distribution of PCBs in the Lake Michigan ecosystem, transport

TABLE 1.22 Estimated PCB Levels in U.S. Coastal Sediments

	Sediment Volume (km^3)	PCB Concentration (μg/kg)		Amount of PCBs (x 10^3 kg)	
		low	high	low	high
Pacific Coast	18	10	50	24	120
Atlantic Coast	30	10	40	590	2400
Gulf of Mexico	4	5	25	20	98
			Total	634	2618

NOTE: The following assumptions are made - the sediment has a mass density of 2. The PCBs are stored in the upper 5 cm of the sediment layer within a distance of 30 km from the shoreline for Atlantic and Gulf of Mexico and 2 km for the Pacific Coast.

mechanisms for removing hydrophobic contaminants were studied (Haile 1977). A fractionation of the PCB residues in the water indicated that nearly 60 percent were associated with heavy particles. The mean settling flux was estimated to be 0.25×10^{-4} g/m^2-yr. Dated sediment samples from the Santa Barbara Basin in California, showed that deposition of PCBs began about 1945 (Hom et al. 1974). The rate of deposition in this study was calculated to be 1.2×10^{-4} g/m^2-yr. PCB deposition rates in sediment from the northwestern Mediterranean region had been estimated also (Elder and Fowler 1977). These calculations were based on PCB contents in a series of cores taken at a water depth of about 400 m. The values ranged from 0.8×10^{-4} g/m^2-yr to 1.2×10^{-4} g/m^2-yr over a period of 15 to 20 years. Fecal pellets from Eupausiids were found to contain PCB residues in the range of 4,800 μg/kg to 38,000 μg/kg dry weight, and it was suggested that sinking zooplankton fecal pellets could have contributed to deposition of PCB compounds in deep ocean sediment.

The results of these studies are in remarkably close agreement, considering the different techniques employed. The findings suggest that PCB deposition rates are similar in both lakes and marine coastal areas. Based on these data, a mean deposition rate of 1.0×10^{-4} g/m^2-yr is proposed.

Rate of Release of PCBs Several studies indicate that PCBs may be released from the site of deposition in the sediment by biotic or abiotic processes. After such release, PCBs can be incorporated in the food web or reintroduced into the water column (Veith and Comstock 1975, Young et al. 1977, Courtney and Langstone 1978, Olsson et al. 1978, Sodergren and Larssom 1979, and Weiniger 1978). However, the results reported are only qualitative and do not provide estimates of the quantity being released.

LITHOSPHERIC COMPARTMENT

The estimated distribution and amount of PCBs in the lithosphere are summarized in Table 1.23. The data used to derive the estimates are discussed in the following subsections. The uncertainty involved in attempting to estimate the distribution and amount of PCBs in the

TABLE 1.23 Estimated PCB Levels in the Lithosphere

Compartment	Biomass (kg)	PCB Concentration (mg/kg) low	high	Amount of PCBs (kg) low	high
Soil (1 cm)	1.35×10^{14}	2×10^{-7}	2×10^{-3}	2.7×10	2.7×10^{5}
Plants	6.35×10^{13}	2×10^{-3}	4×10^{-2}	1.3×10^{5}	2.5×10^{6}
Wildlife	6.35×10^{8}	2×10^{-2}	4×10^{-1}	1.3×10^{1}	2.6×10^{2}
Livestock	6.35×10^{10}	2×10^{-3}	10^{-1}	1.3×10^{2}	6.3×10^{3}
Man	1.4×10^{10}	3.5×10^{-1}		4.9×10^{3}	4.9×10^{3}
			Total	1.4×10^{5}	2.8×10^{6}

biota must be noted, particularly in the two largest compartments, soil and plants.

PCBs in Soil

Little information is available on the concentration of PCBs in soil. These compounds generally are not detectable in agricultural soil. Although many "hot spots" with high PCB levels exist in some industrialized locations, the estimated average soil concentration for metropolitan areas in the United States is about 0.002 mg/kg (Carey and Gowen 1976). In Japan, the average concentration is 0.08 mg/kg (Fujiwara 1975).

Whereas the pattern of PCB pollution in lakes appears to be controlled by atmospheric fallout, results of a nationwide network of sampling stations indicate that PCB concentrations in stream sediment are the result of accumulation downstream from known point sources of PCB discharge or from industrialized urban areas (STORET-U.S. EPA Water Quality Analysis Branch, Monitoring and Data Support Division). The contrast in patterns suggests that PCBs in rainfall make negligible contributions to sediment in streams. A possible explanation for this is that the primary source of water in stream flow is water that seeps out of the ground (defined as the base flow of streams). Only rural contributions to streams come via rain that falls on the water in the channel. (Appendix A, Table A.1, shows that the aggregate surface area of streams is only one quarter that of lakes and large reservoirs.) The inferred lack of PCBs from groundwater discharge is consistent with data from New York State indicating negligible losses from landfill to ground water (Leis and Metry 1978, Hetling et al. 1978). But if PCBs are widely dispersed in rainfall yet virtually absent from ground water, one must conclude that PCBs are accumulating in the soil and near-surface sediment underlying the soil. As already noted, data are scarce, therefore it appears to be important to monitor PCB levels in soil to find out how much enters from rainfall and what their fate is after entering.

The compartmental biomass and estimated PCB concentrations are presented in Table 1.23. Using these values plus a land area of 9 x 10^6 km^2, soil depth of 1 cm and soil density of 1.5 g/cm^3, the range of PCB burden was calculated as 2.70 x 10 kg to 2.7 x 10^5 kg.

Plants

An average PCB concentration in vegetation has been reported only for remote areas of Germany (0.02 mg/kg, from Klein and Weisgerber 1976). The high and low estimates in Table 1.23 were based on ranges given in the German study and represent mg/kg fresh weight. The biomass for plants was estimated in relation to livestock biomass (6.35 x 10^{13} kg, from Robinson 1975), and the PCB burden was calculated as 1.3 x 10^5 kg to 2.5 x 10^6 kg.

Wildlife

Most of the available data on PCB concentrations found in birds deal with predators. The level in tissue samples generally depends on the type of diet for each species studied. Avian species feeding mainly on small birds and mammals have PCB concentrations in liver samples that range from nondetectable to 50 mg/kg fresh weight; those feeding on fish have concentrations ranging from nondetectable to 900 mg/kg fresh weight (Prestt et al. 1970). The data indicate that liver concentrations are approximately 1.5 to 2 times the concentrations found in whole body samples. Residues in birds that feed on worms and insects are considerably lower compared to the above values. PCB tissue levels of 0.55 mg/kg fresh weight have been reported for starlings (Sodergren and Ulfstrand 1972), and 0.1 mg/kg to 0.65 mg/kg fresh weight reported for robins (White 1976). PCB levels in tissue of herbivorous birds average 0.2 mg/kg fresh weight (calculated for mourning doves, Kreitzer 1974).

Little information is available on PCB levels in wild mammals. Mice in Iceland contained tissue levels of 0.06 mg/kg fresh weight (Bengtsson and Sodergren 1974). Lipid samples from foxes and polar bears in Greenland contained 2.8 mg/kg and 21 mg/kg of PCBs, respectively (Clausen and Berg 1975).

PCB levels in wildlife were investigated during an extensive study of the Hudson River. Muscle and liver samples from snapping turtles in the Hudson Estuary had concentrations ranging from 0.05 μg/kg to 174 μg/kg (Hetling et al. 1978). Samples of liver tissue of selected animals indicated PCB ranges of nondetectable to 0.26 μg/kg in raccoons, 0.18 μg/kg to 0.90 μg/kg in mink and 0.44 μg/kg to 1.70 μg/kg in otters (New York State Department of Environmental Conservation 1978).

It is difficult to derive an average PCB concentration for wildlife. For the purpose of Table 1.23, low and high PCB concentrations are estimated as 0.02 mg/kg and 0.4 mg/kg fresh weight, and the biomass value was estimated in relation to livestock biomass (6.35×10^8 kg, Robinson 1975). The range for PCB burden in wildlife is calculated to be 1.3×10^1 kg to 2.6×10^2 kg.

Livestock

Levels of PCBs in livestock may be derived from data on PCB concentrations in meat. In Japan, average concentrations in wet weight are 0.02 mg/kg in beef, 0.05 mg/kg in pork, 0.004 mg/kg in mutton, and 0.06 mg/kg in chicken (Fujiwara 1975). Because similar values have been reported from other countries, a concentration range of 0.002 mg/kg to 0.1 mg/kg was used for Table 1.23. The biomass was estimated using USDA *Agricultural Statistics 1972 and 1974* for the number and average weight of livestock on U.S. farms. The estimated burden for this compartment ranges from 1.3×10^2 kg to 6.3×10^3 kg.

PCBs in Humans

Levels of PCBs in human adipose tissue have been well documented. A summary of the data is presented in Table 1.24. In those cases in which the original data were not presented in mg/kg of lipid, the values were recalculated assuming that adipose tissue contains 76 percent lipid.

The concentrations detected in adipose tissue can be converted into whole body levels, assuming that lipid constitutes 30 percent of body weight. Accordingly, a PCB concentration of 0.35 mg/kg body weight has been suggested. A population size of 2×10^8 with an average weight of 70 kg per person was used for calculating biomass (1.4×10^{10} kg) in Table 1.23. The PCB burden associated with humans is 4.9×10^3 kg.

Transport Between Lithosphere and Atmosphere

Volatilization rates for PCBs appear to depend on experimental conditions. A loss of 40 to 50 percent was observed for Aroclor 1254 in either dry or wet sand, at 26°C over a period of 4 weeks. In contrast, loss of Aroclor 1254 from a soil containing 3.1 percent organic

TABLE 1.24 PCBs in Human Adipose Tissue

Country	Mean Concentration (mg/kg)
Denmark[1]	5.0
Japan[2]	4.7
West Germany[3]	7.9
Austria[1]	4.6
U.K.[1]	1.3
Norway[1]	0.9
Canada[4]	1.2
East Germany[5]	6.4
Israel[5]	3.6
New Zealand[6]	0.9
U.S.A.[7]	1.2

SOURCE:
[1]Modified from Kraul and Karlog (1976).
[2]Modified from Fujiwara (1975).
[3]Modified from Acker and Schulte (1974).
[4]Modified from Mes et al. (1977).
[5]Modified from Wassermann et al. (1974).
[6]Modified from Solly and Shanks (1974).
[7]Modified from Kutz and Strassman (1976).

matter was regarded as negligible. PCBs in other soils with lower organic content may undergo some volatilization. When allowance is made for differences in experimental conditions, the rate of vaporization of Aroclor 1254 from sand appears to be of the same order of magnitude as the volatilization loss of Aroclor 1254 from itself which is 2.0 x 10^{-6} g/cm^2-day at 26°C (Haque et al. 1974). If a linear relationship is assumed between PCB concentration in the top 1 cm of soil and the rate of vaporization, the expected vaporization rate from a typical sandy soil concentration of 2 μg/kg (Carey and Gowen 1976) would be 4 μg/m^2-day, or on the same order as the atmospheric deposition rate in the Great Lakes area (Murphy and Rzeszutko 1977).

The results indicate that PCBs deposited by atmospheric fallout onto sand, and presumably other inorganic materials (such as rock or concrete), will not accumulate but will reenter the atmosphere via volatilization at a rapid rate (2.0 μg/cm^2-day). PCBs deposited onto high-organic soils will reenter the atmosphere at a slower rate. However, the low concentrations observed in soils (Carey and Gowen 1976) combined with the estimated rate of deposition (Table 1.4) indicate that the residence time would be on the order of one month or less.

High-organic soils in the United States are typically covered--at least partially--with vegetation. Foliage and grasses may be efficient collectors of those PCBs associated with either particulate matter or precipitation fallout, and the vegetation probably would not easily release PCBs via volatilization; this assumption is supported by data from a study in New York. PCB levels were measured in or on foliage samples taken at control sites, upwind of facilities that formerly used PCBs. Ranges of 0.2 μg/kg to 0.8 μg/kg were reported (Leis and Metry 1978). PCBs would be introduced to the upper soil layer through leaf litter decomposition processes, but the average concentration in soil would be diluted by the additional presence of noncontaminated plant materials.

PCBs deposited on agricultural lands may never accumulate in the top layer of soil if the food crop is removed totally. Lands supporting row crops, grains, and other similar plants are typically turned under yearly, so that the PCBs are spread throughout a greater depth than the 1 cm mentioned above. Crop foliage and pasture growth typically are removed and used as food or bedding for livestock.

Until more experiments are conducted studying processes of transfer between the atmosphere and the lithosphere, many questions regarding the net flux between these major environmental compartments cannot be answered.

DISTRIBUTION MODEL FOR PCBs FROM DOMESTIC SOURCES

This section of the report uses estimates of domestic PCB usage from 1930 to 1977 to derive a Mobile Environmental Reservoir (MER). The MER is defined as that portion of PCBs existing in a form that can be accumulated in the environmental media and biota, including amounts detected in air, water, sediment, upper levels of soil, and biota trophic levels. Material in landfills is specifically excluded as PCBs within these sites are considered generally to be immobile.

The composition of the MER derived from data on domestic production and use, is categorized by chlorine content of the biphenyl molecule. These categories are described below in descending order of biodegradability:

- Category I: PCBs containing one to three chlorine atoms per molecule (mono-, di-, tri-isomers).
- Category II: PCBs containing four chlorine atoms per molecule (tetra-isomers).
- Category III: PCBs containing five or more chlorine atoms per molecule (penta- through deca-isomers).

Magnitude and Composition of the MER

Data on PCB production and sales, including information on domestic sales by category of use and Aroclor designation, were obtained for the period 1957 to 1974 (Durfee et al. 1976). Using these data and known distributions of chlorine atoms per molecule for each Aroclor (Hutzinger et al. 1974) the yearly domestic sales can be calculated for each of three categories over the specified period.

The available data were extended to cover the entire period through 1977 by deriving best-fit empirical functions from Categories I, II and III of PCBs used in both electrical equipment (closed systems) and nonelectrical applications (open-end and nominally closed systems). The resulting equations are:

Electrical Applications (units are kg/yr)

Category I Usage* $= 2.4 \times 10^5 e^{0.097t}$ (t = 0 at 1930)
Category II Usage $= 1.16 \times 10^5 e^{0.098t}$ (t = 0 at 1930)
Category III Usage $= 2.24 \times 10^5 e^{0.102t}$ (t = 0 at 1930)

*In order to establish a zero base (i.e. start of production), subtract 3×10^6 kg.

Nonelectrical Applications (kg/yr)

Category I Usage $= 2.43 \times 10^5 e^{0.177t}$ (t = 0 at 1950)
Category II Usage $= 2.21 \times 10^5 e^{0.155t}$ (t = 0 at 1959)
Category III Usage $= 7.35 \times 10^4 e^{0.105t}$ (t = 0 at 1950)

In these equations, time (t) is measured in years from the initial dates of production (t = 0) as specified for each equation. Nonelectrical applications became significant around 1950 and correlates approximately with commercialization of Aroclor 1248. After 1971, sales for nonelectrical applications effectively became zero.

The model described above results in a total cumulative use of PCBs from domestic sources through 1975 of 6.1×10^8 kg, 7 percent higher than the value of 5.7×10^8 kg previously estimated (Durfee et al. 1976). The production percentages, extended through 1977 for each of the three PCB categories, are:

Category I: 38 percent
Category II: 23 percent
Category III: 39 percent

The assumed loss factors for production and use of PCBs summarized in Table 1.25 are based on engineering judgment applied to the available information (Durfee et al. 1976). The factors are assumed to apply to all categories, as noted, through the cessation of domestic production in 1977.

Table 1.26 summarizes the cumulative losses to the MER calculated by applying loss factors from Table 1.25 to the production-use model above. For Category I cumulative losses in electrical use, the basic equation

TABLE 1.25 Assumed Loss Factors for Production and Usage of PCBs (1930-1977)

Function	Percent of Overall Loss Factor	Percent of Fraction Entering the MER
Primary Production	2[a]	50
Transport/Handling	0.5[a]	100
First Tier Usage	2[b]	50
In-Service Electrical Equipment	0.1 of material in service per year[c]	100
Material in Nonelectrical Applications	10 of material in service per year[d]	100

[a]Applicable to total domestic production.
[b]Applicable to material used in each of two general applications (electrical and nonelectrical).
[c]Useful service life of 40 years.
[d]Useful service life of 6 years.

without zero correction was used, and the result corrected by subtracting 0.2 x 10^7 kg. A similar zero correction, by subtracting 0.05 x 10^7 kg, is required for Category II losses from electrical service. These corrections are required because of the extrapolations made for the period from 1930 to 1957; however, the factors are small when compared to the total cumulative amounts in the MER for each category.

The results in Table 1.26 indicate that, as of 1977, a total of about 8.25 x 10^7 kg of PCBs had entered the MER. The amounts by PCB category were:

	Amounts (x 10^7 kg)
Category I:	3.12
Category II:	1.83
Category III:	3.30

The estimate of 8.25 x 10^7 kg is directly comparable with the 1972 estimate of 8.2 x 10^7 kg (Nisbet and Sarofim 1972), as well as the value derived in 1975 for all PCBs available to the biota (ignoring degradation processes) of 7.8 x 10^7 kg (Durfee et al. 1976). The MER model suggests that the cumulative amounts of PCBs

TABLE 1.26 Estimated Cumulative Losses of PCBs to the Mobile Environmental Reservoir as of 1977 by PCB Category (Chlorine Content) and Production/Use Function

	PCB Losses to the MER as of 1977 (kg x 10^7)			
Production/Use Function	Category I	Category II	Category III	Total
Primary Production	0.30	0.18	0.31	0.79
Transportation/Handling	0.15	0.09	0.16	0.40
First Tier Usage	0.37	0.23	0.38	0.98
In-Service Elec. Equip.	0.14*	0.13*	0.19	0.46
Non-Electrical Applications	2.16	1.20	2.26	5.62
	3.12	1.83	3.30	8.25

*Zero-corrected (see text).

in the mobile environment as of 1972 and 1975 were significantly less than previously estimated.

Removal of PCBs from the MER

The major identified process for removal of PCBs from the MER (as defined above) is land disposal of sewage sludge. Using the estimate of 2.4×10^5 kg/yr of PCBs in sewage sludge (see the subsection on "Industrial and Municipal Effluents" above), and applying this value over a 20-year period, it is estimated that about 4.8×10^6 kg of PCBs have been removed from the MER via sewage treatment and subsequent sludge disposal.

SUMMARY COMPARISON OF INVENTORY

The environmental inventory is summarized in Table 1.27. The total estimated range of domestically derived PCBs in or removed from the environment is between 1.3×10^7 kg and 8.4×10^7 kg. These values compare favorably to the estimated MER of 8.25×10^7 kg.

The major portion of the environmentally available PCBs is contained in marine waters, accounting for 50 to 80 percent of the total inventory. The total amount of PCBs in the North American Basin is estimated to be 0.6×10^7 kg to 6.6×10^7 kg. In comparison to these values, water concentrations from the Pacific Ocean and Gulf of Mexico attributable to domestically produced PCBs can be considered negligible. It should be noted that the values in Table 1.27 are based on measurements referring mainly to PCBs in Category III and that the high estimate is greater than the estimate of environmental releases of Category III materials. It is possible that some of the oceanic PCBs originate from European as well as North American sources. This compartment appears to be a dominant reservoir for PCBs even if the oceanic inventory is high by a factor of two. The rate at which PCBs will eventually reach (and perhaps become bound to) deep ocean sediment, however, cannot be estimated on the basis of current knowledge.

The close agreement between environmental measurements and the results of the model obtained in this analysis should not be interpreted as an indication that biodegradation and other processes of destruction of PCBs do not occur in the environment. Biodegradation

TABLE 1.27 Summary of Environmental Inventory of U.S.-Derived PCBs

Compartment	Estimate* (kg) Low		High
Air		1.8×10^4	
Lithosphere	1.4×10^5		2.8×10^6
Fresh Water Sediments	1.4×10^6		7.1×10^6
Fresh Waters	1.2×10^4		3.5×10^4
Fresh Water Biota		3.0×10^4	
Oceanic Sediments	6.6×10^5		2.7×10^6
Oceanic Waters	6.0×10^6		6.6×10^7**
Oceanic Biota		3.0×10^4	
Sewage Sludge		4.8×10^6	
Total	1.3×10^7		8.4×10^7
Percentage of Total in Oceanic Waters	50%		79%

*Where only one estimate was made it is placed between the low and high columns; only two significant digits are used for the estimates and totals.

**Some PCBs in this estimate may be derived from European sources.

may be the ultimate fate of the PCBs now in the MER. The results, however, do indicate that biodegradation (or other processes resulting in chemical alteration of the molecules) has not destroyed a major fraction of environmental PCBs at this time.

RECOMMENDATIONS

1. Because the North Atlantic Ocean appears to be the ultimate sink, further studies are needed to determine PCB concentrations and distribution in the marine environment.

2. Although modeling was useful in assessing large-scale distribution and fate of toxic substances, more adequate data bases are needed to develop transport models of environmental pollutants. For example,

further research is needed to determine the processes of transfer between the environmental compartments.

3. EPA should continue to develop better analytical techniques, particularly those enabling assessment of the extent of PCB fractionation. Because of continuing analytical problems it is difficult to determine the extent of ecosystem contamination. For aquatic systems, the most effective indicator of both the degree of PCB pollution and the potential hazards to humans and wildlife is the PCB content of fish. To effectively use this indicator, factors such as age, species, and nutritional habits of fish must be considered. Detailed information on the individual PCB isomers should also be reported together with the specific Aroclor designation.

4. Freshwater sediments are important continental sinks for PCBs and special attention should be given to monitoring programs and studies of PCB cycling from sediment through human food chains.

REFERENCES

Acker, L. and E. Schulte (1974) Chlorkohlenwasserstoffe im Menschlichen fett. Naturwissenschaften 61:32.

Ahmed, M. and D.D. Focht (1973) Degradation of PCBs by two species of Achromobacter. Can. J. Microbiol. 19:47-52.

Bengtsson, S.A. and A. Sodergren (1974) DDT and PCB in airborne fallout and animals in Iceland. Ambio 3(2):84-86.

Berg, A.K. and R.S. Peoples (1977) Distribution of PCBs in a municipal wastewater treatment plant and environs. Sci. Total Environ. 8:197-204.

Bidleman, T.F. and C.E. Olney (1974) Chlorinated hydrocarbons in the Saragasso Sea atmosphere and surface water. Science 183:516-518.

Bidleman, T.F., C.P. Rice, and C.E. Olney (1978) High molecular weight chlorinated hydrocarbons in the air and sea: Rates and mechanisms of air/sea transfer. Pages 323-351, Marine Pollutant Transfer. Edited by H.L. Windom and R.A. Duce. Lexington, Mass.: D.C. Heath and Company.

Carey, A.E. and J.A. Gowen (1976) PCBs in agricultural and urban soil. Pages 195-198, Proceedings of National Conference on Polychlorinated Biphenyls, November 19-21, 1975, Chicago, Illinois. Report No.

EPA-560/6-75-004. Washington, D.C.: U.S. Environmental Protection Agency.

Carr, R.A., G.Y. Contos, R.L. Durfee, V.C. Fong, and E.G. McKay II (1977) Involvement of PCBs in the Pulp and Paper Industry. Office of Toxic Substances. Final Task Report No. EPA 560/6-77-005, Contract No. 68-01-3259. Washington, D.C.: U.S. Environmental Protection Agency.

Clark, W.D. and L.C. Murdock (1972) Chester River Study: A Joint Investigation by the State of Maryland Department of Natural Resources and Westinghouse Electric Corporation, Vol. I. Annapolis, Md.: Westinghouse Electric Oceanic Division.

Clausen, J. and O. Berg (1975) The content of polychlorinated hydrocarbons in Artic ecosystems. Pure Appl. Chem. 42:223.

Council on Environmental Quality (1970) Environmental Quality. The First Annual Report. Washington, D.C.: U.S. Government Printing Office.

Courtney, W.A.M. and W.J. Langstone (1978) Uptake of PCB (Aroclor 1254) from sediment and from sea water in two intertidal polychaetes. Environ. Pollut. 15:303-309.

Crump-Wiesner, H.J., H.R. Feltz, and M.L. Yates (1973) A study of the distribution of polychlorinated biphenyls in the aquatic environment. J. Res. USGS 1(5):603-607.

Dawson, R. and J.P. Riley (1977) Chlorine-containing pesticides and polychlorinated biphenyls in British coastal waters. Estuarine Coastal Mar. Sci. 5:55-69.

de Lappe, B.W., R.W. Risebrough, A.M. Springer, T.T. Schmidt, J.C. Sharopshire, E.F. Letterman, and J.R. Payne (In press) Sampling and measurement of hydrocarbons in natural waters. Proceedings of International Symposium on Analysis of Hydrocarbons. Edited by B.F. Afghan and D. Mackay.

Dennis, D.S. (1976) Polychlorinated biphenyls in the surface waters and bottom sediments of the major drainage basins of the United States. Pages 183-194, Proceedings of National Conference on Polychlorinated Biphenyls, November 19-21, 1975, Chicago, Illinois. Report No. EPA-560/6-75-004. Washington, D.C.: U.S. Environmental Protection Agency.

Dexter, R.N. and S.P. Pavlou (1978) Mass solubility and aqueous activity coefficients of stable organic chemicals in the marine environment: PCBs. Mar. Chem. 6:41-53.

Doskey, P.V. (1978) Transport of Airborne PCBs to Lake Michigan. M.S. Thesis, Water Chemistry Program. Madison, Wis.: University of Wisconsin.

Duce, R.A. and E.K. Duursma (1977) Inputs of organic matter to the ocean. Mar. Chem. 5:319-339.

Durfee, R.L., G. Contos, F.C. Whitmore, J.D. Barden, E.E. Hackman III, and R.A. Westin (1976) PCBs in the United States--Industrial Use and Environmental Distribution. Report No. EPA 56016-16-005. Springfield, Va.: Versar, Inc.

Eder, G. (1976) PCBs and compounds of the DDT group in sediments of the central North Sea and the Norwegian Depression. Chemosphere 2:101-106.

Edgington, D.N. and J.A. Robbins (1976) Records of lead deposition in Lake Michigan sediments since 1800. Environ. Sci. Technol. 10:266-273.

Eisenreich, S.J., G.J. Hollod and T.C. Johnson (1979) Accumulation of PCBs in Surficial Lake Superior Sediments: Atmospheric Deposition. A report of the Limnological Research Center, University of Minnesota, Minneapolis, Minnesota.

Elder, D.L. and S.W. Fowler (1977) PCBs: Penetration into the deep ocean by zooplankton fecal pellets transport. Science 197:459-461.

Fujiwara, K. (1975) Environmental and food contamination with PCBs in Japan. Sci. Total Environ. 4:219.

Fuller, B., J. Gordon, and M. Kornreich (1977) Environmental Assessment of PCBs in the Atmosphere. EPA Report No. EPA-450/3-77-045. McLean, Va.: The Mitre Corporation.

Furukawa, K., K. Tonomura, and A. Kamibayashi (1978) Effect of chlorine substitution on the biodegradability of polychlorinated biphenyls. Appl. Environ. Microbiol. 35:223-227.

Giam, C.S., H.S. Chan, and G.S. Neff (1976) Concentrations and fluxes of phthalates, DDTs and PCBs to the Gulf of Mexico. In, Marine Pollutant Transfer. Edited by H. Windom and R. Duce. Lexington, Mass.: D.C. Heath and Company.

Glooschenko, W.A., W.M.J. Strachan, and R.C. Sampson (1976) Distribution of pesticides and PCBs in water, sediments, and seston of the Upper Great Lakes. Pestic. Monit. J. 10:61-67.

Graham, J.M. (1976) Levels of PCBs in Canadian commercial fish species. Pages 155-160, Proceedings of National Conference on Polychlorinated Biphenyls, November 19-21, 1975, Chicago, Illinois. Report No. EPA-560/6-75-004. Washington, D.C.: U.S. Environmental Protection Agency.

Greichus, Y.A., A. Greichus, B.A. Ammann, D.J. Call, D.C.D. Hamman, and R.M. Pott (1977) Insecticides, polychlorinated biphenyls and metals in African Lake ecosytems. I. Hartbeespoort Dam, Transvaal and Voelvlei Dam, Cape Province, Republic of South Africa. Arch. Environ. Contam. Toxicol. 6:371-383.

Greichus, Y.A., A. Greichus, B.D. Ammann, and J. Hopcraft (1978a) Insecticides, polychlorinated biphenyls and metals in African Lake ecosystems. III. Lake Nakuru, Kenya. Bull. Environ. Contam. Toxicol. 7:454-461.

Greichus, Y.A., A. Greichus, H.A. Draayer, and B. Marshall (1978b) Insecticides, PCBs and metals in African Lake ecosystems. II. Lake Mellwaine, Rhodesia. Bull. Environ. Contam. Toxicol. 7:444-453.

Haile, C.L. (1977) Chlorinated Hydrocarbons in the Lake Ontario and Lake Michigan Ecosystems. Thesis. Madison, Wis.: University of Wisconsin.

Hakansson, L. (1977) Sediments as indicators of contamination - investigations in the four largest Swedish Lakes. Report No. SNV PM 839. Environmental Protection Board of Sweden.

Halter, M.T. and H.E. Johnson (1977) A model system to study the desorption and biological availability of PCB in hydrosoils. Pages 178-195, Aquatic Toxicology and Hazard Evaluation, ASTM STP 634. Edited by F.L. Mayer and J.L. Hamelink. Philadelphia: American Society for Testing and Materials.

Hamelink, J.L., R.C. Waybrant, and R.C. Ball (1971) A proposal: exchange equilibria control the degree chlorinated hydrocarbons are biologically magnified in lentic environments. Trans. Am. Fish Soc. 100:207.

Haque, R., D.W. Schmedding, and V.H. Freed (1974) Aqueous solubility, adsorption, and vapor behavior of polychlorinated biphenyl Aroclor 1254. Environ. Sci. Technol. 8:139-142.

Haque, R. and D.W. Schmedding (1975) The method of measuring the water solubility of hydrophobic chemicals: solubility of five polychlorinated biphenyls. Bull. Environ. Contam. Toxicol. 14:13-18.

Hargrave, B.T. and G.A. Phillips (1975) The lack of accumulation of DDT residue by benthic invertebrates and demersal fish in St. Margaret's Bay, Nova Scotia. Int. Counc. Expl. Sea, CM-E:32.

Harvey, G.R. and W.G. Steinhauer (1974) Atmospheric transport of polychlorobiphenyls to the North Atlantic. Atmos. Environ. 8:777-782.

Harvey, G.R. and W.G. Steinhauer (1976a) Biogeochemistry of PCB and DDT in the North Atlantic, Vol. I. Pages 203-221, Environmental Biogeochemistry. Edited by J.O. Nriagu. Ann Arbor, Mich.: Ann Arbor Science Publishers, Inc.

Harvey, G.R. and W.G. Steinhauer (1976b) Transport pathways of polychlorinated biphenyls in Atlantic water. J. Mar. Res. 34:561-575.

Harvey, G.R., W.G. Steinhauer, and H.P. Miklas (1974) Decline of PCB concentrations in North Atlantic surface water. Nature 252:387-388.

Hetling, L., E. Horn, and T.J. Tofflemire (1978) Summary of Hudson River PCB Study Results. Technical Report No. 51. Albany, N.Y.: New York State Department of Environmental Conservation.

Hom, W., R.W. Risebrough, A. Soutar, and D.R. Young (1974) Deposition of DDE and PCB in dated sediments of the Santa Barbara Basin. Science 184:1197-1199.

Hutzinger, O., S. Safe, and V. Zitko (1974) The Chemistry of PCBs. Cleveland, Ohio: CRC Press.

Junge, C.E. (1977) Basic considerations about trace constituents in the atmosphere as related to the fate of global pollutants. Pages 7-25, Part 1--The Mechanisms of Interaction between Environments and Mathematical Modeling and the Physical Fate of Pollutants. Edited by I.H. Suffet. New York: Wiley-Interscience.

Klein, W. and I. Weisgerber (1976) PCBs and environmental contamination. Environ. Qual. Saf. 5:237.

Kleinert, S.J. (1976) The PCB Problem in Wisconsin. Report for the Joint Hearing of the Assembly of Environmental Quality Committee and Senate and Assembly of Natural Resources Committees on HR 212 Administrative Rules, Madison, Wisconsin, September 21.

Kraul, I. and O. Karlog (1976) Persistent organochlorinated compounds in human organs collected in Denmark 1972-73. Acta Pharmacol. Toxicol. 38:38.

Kreitzer, J.F. (1974) Residues of organochlorine pesticides, mercury and PCBs in mourning doves from

Eastern United States - 1970-71. Pestic. Monit. J. 7:195.
Kutz, F.W. and S.C. Strassman (1976) Residues of polychlorinated biphenyls in the general population of the United States. Page 139, Proceedings of National Conference on Polychlorinated Biphenyls, November 19-21, 1975, Chicago, Illinois. Report No. EPA 560/6-75-004. Washington, D.C.: U.S. Environmental Protection Agency.
Law, L.M. and D.F. Goerlitz (1974) Selected chlorinated hydrocarbons in bottom materials from streams tributary to San Francisco Bay. Pestic. Monit. J. 8:33-36.
Leis, W.M. and A. Metry (1978) Study of Migration of PCBs from Landfills and Dredge Spoil Sites and Related to the Hudson River. Final Report submitted by Weston Environmental Consultants-Designers, West Chester, Pennsylvania to New York State Department of Environmental Conservation, Albany, New York.
Leonard, J. (1977) Organohalogens in Coastal Sediments from the Maritime Provinces, Canada. Bedford Institute of Oceanography Report Series, June. Halifax, Nova Scotia: Bedford Institute of Oceanography.
Linko, R.R., P. Rantamaki, and K. Urpo (1974) PCB residues in plankton and sediments in the southwestern coast of Finland. Bull. Environ. Contam. Toxicol. 12:733-738.
Lunde, F., J. Gether, N. Gjos, and N.S. Lande (1977) Organic micro-pollutants in precipitation in Norway. Atmos. Environ. 11:1007-1014.
Mackay, D. and P.J. Leinonen (1975) Route of evaporation of low-solubility contaminants from water bodies to the atmosphere. Environ. Sci. Technol. 9:1178-1181.
Mackay, D. and A.W. Wolkoff (1973) Rate of evaporation of low-solubility contaminants from water bodies to atmosphere. Environ. Sci. Technol. 7:611-614.
McClure, E.V. and J. Lagrange (1977) Deposition of heavy chlorinated hydrocarbons from the atmosphere. Bull. Environ. Contam. Toxicol. 17(2):219-224.
Mes, J., D.S. Campbell, R.N. Robinson, and D.J.A. Davies (1977) Polychlorinated biphenyls and organochlorine pesticide residues in adipose tissue of Canadians. Bull. Environ. Contam. Toxicol. 17:196.
Mieure, J.P., O. Sicks, R.G. Kaley, B.W. Saeger (1975) Characterization of polychlorinated biphenyls. Pages 84-93, Proceedings of National Conference on

Polychlorinated Biphenyls, November 19-21, 1975, Chicago, Illinois. Report No. EPA 560/6-75-004. Washington, D.C.: U.S. Environmental Protection Agency.

Moein, G.J. (1976) Study of Distribution and Fate of PCBs and Benzenes After Spill of Transformer Fluid. Report No. EPA 904/9-76-014. Washington, D.C.: U.S. Environmental Protection Agency.

Moein, G.J., A.J. Smith, and P.L. Stewart (1976) Follow-up study of the distribution and fate of PCBs and benzenes in soil and ground water samples after an accidental spill of transformer fluid. Pages 368-372, Proceedings of 1976 National Conference on Control of Hazardous Materials Spills. Rockville, Md.: Information Transfer, Inc.

Murphy, T.J. and C.P. Rzeszutko (1977) Precipitation inputs of PCBs to Lake Michigan. J. Great Lakes Res. 3(3-4):305-312.

Nadeau, R.J. and R.P. Davies (1974) Investigation of PCB in the Hudson River. Washington, D.C.: U.S. Environmental Protection Agency.

National Research Council (1977) An Assessment of Mercury in the Environment. Committee for Scientific and Technical Assessments of Environmental Pollutants, Environmental Studies Board, Commission on Natural Resources. Washington, D.C.: National Academy of Sciences.

New York State Department of Environmental Conservation (1978) Toxic Substances Impacting on Fish and Wildlife. Division Fish and Wildlife Monthly Report 12, March 20. Albany, N.Y.: New York State Department of Environmental Conservation.

Nisbet, I.C.T. (1976) Criteria Document for PCBs. Report No. PB-225 397. Washington, D.C.: U.S. Department of Commerce.

Nisbet, I.C.T. and A.F. Sarofim (1972) Rates and routes of transport of PCBs in the environment. Environ. Health Perspect. 1:21-38.

Oden, S. and J. Ekstedt (1976) PCB and DDT in Baltic sediments. Ambio Special Report 4:125-129.

Oloffs, P.C., L.J. Albright, S.Y. Szeto, and J. Lau (1973) Factor affecting the behavior of five chlorinated hydrocarbons in two natural waters and their sediments. J. Fish. Res. Bd. Can. 30:1619-1623.

Olsson, M., S. Jensen, and L. Reutergard (1978) Seasonal variation of PCB levels in fish - An important factor in planning monitoring programs. Ambio 7:66-69.

PLUARG-International Reference Group on Great Lakes Pollution from Land Use Activities (1978) Environmental Management Strategy for the Great Lakes System. Final Report to the International Joint Commission. Windsor, Ontario, Canada.

Prestt, I., D.J. Jeffries, and N.W. Moore (1970) Polychlorinated biphenyls in wild birds in Britain and their avian toxicity. Environ. Pollut. 1:3.

Risebrough, R.W., B.W. de Lappe, and W. Walker II (1976) Transfer of higher molecular weight chlorinated hydrocarbons to the marine environment. Pages 261-321, Marine Pollutant Transfer. Edited by H.L. Windom and R.A. Duce. Lexington, Mass.: D.C. Heath and Company.

Risebrough, R.W., V. Freeland, G.R. Harvey, H.P. Miklas, and G.M. Carmigan (1972) PCB residues in Atlantic zooplankton. Bull. Environ. Contam. Toxicol. 8:345-355.

Robinson, J. (1975) Mathematical models in ecochemistry. Pure Appl. Chem. 42:139-153.

Sanders, J.E. (1979) Sedimentologic Aspects of the PCB Pollution of the Hudson River. (Manuscript in preparation)

Sargent, D.H., R.A. Carr, and G.A. Vogel (1977) PCBs Removal in Publicly-Owned Treatment Works. Final Report, Contract No. 68-01-3273, Task 13. Criteria and Standards Division. Washington, D.C.: U.S. Environmental Protection Agency.

Sayler, G.S. and R.R. Colwell (1976) Partitioning of mercury and PCB by oil, water, and suspended sediment. Environ. Sci. Technol. 10:1142-1145.

Slinn, W.G.N., L. Hasse, B.B. Hicks, A.W. Hogan, D. Lal, P.S. Liss, K.O. Munnich, G.A. Sehmel, and O. Vittori (1978) Some aspects of the transfer of atmospheric trace constituents past the air/sea interface. Atmos. Environ. 12(11):2055-2087.

Sodergren, A. (1972) Chlorinated hydrocarbons residues in airborne fallout. Nature 236:395-397.

Sodergren, A. (1973) Transport, distribution and degradation of organochlorine residues in a South Swedish lake ecosystem. Vatten 2:90-108.

Sodergren, A. (1975) Monitoring DDT and PCB in airborne fallout. Environ. Qual. Saf. 3:803-810.

Sodergren, A. and P. Larssom (1979) Transport of PCBs from Lake Sediments to the Atmosphere via the Aquatic Surface Microlayer. (To be published)
Sodergren, A. and P. Ulfstrand (1972) DDT and PCB relocate when caged robins use fat reserves. Ambio 1:36-40.
Solly, S.R.B. and V. Shanks (1974) Polychlorinated biphenyls and organochlorine pesticides in human fat in New Zealand. N.Z. J. Sci. 17:535.
Swain, W.R. (1979) Chlorinated organic residues in fish, water, and precipitation from the vicinity of Isle Royale, Lake Superior. J. Great Lakes Res. (In press)
Travers, I.C. and R.C.H. Wilson (1977) PCBs in the Atlantic Provinces. Surveillance Report No. EPS-5-AR-77-13. Halifax, Nova Scotia: Environmental Protection Service.
Tucker, E.S., V.W. Saeger, and O. Hicks (1975) Activated sludge primary biodegradation of PCB. Bull. Environ. Contam. Toxicol. 14:705-713.
Tulp, M.Th.M., R. Schmitz, and O. Hutzinger (1978) The bacterial metabolism of 4,4' dichlorobiphenyl and its suppression by alternative carbon sources. Chemosphere 7:103-108.
U.S. Environmental Protection Agency (1978) Polychlorinated biphenyls (PCBs): Disposal and marking. Federal Register 43(34):7149-7164.
U.S. Environmental Protection Agency (1976) Review of PCB Levels in the Environment. Report No. EPA-560/7-76-001. Washington, D.C.: U.S. Environmental Protection Agency.
Veith, G.D. (1972) Recent fluctuations of PCBs in the Green Bay Wisconsin region. Environ. Health Perspect. 1:51-52.
Veith, G.D. (1975) Baseline concentrations of polychlorinated biphenyls in Lake Michigan fish, 1971. Pestic. Monit. J. 9:21-29.
Veith, G.D. and V.M. Comstock (1975) Apparatus for continuously saturating water with hydrophobic organic chemicals. J. Fish. Res. Bd. Can. 32:1849-1851.
Veith, G.D., D.W. Kuehl, F.A. Puglisi, G.E. Glass, and J.G. Eaton (1977) Residues of PCBs and DDT in the western Lake Superior ecosystem. Arch. Environ. Contam. Toxicol. 5:487-499.
Versar, Inc. (1978) Gross Annual Discharge to the Waters in 1976 - Polychlorinated Biphenyls. Monitoring and

Data Support Division, Revised Report No. 16, Task 11, Contract No. 68-01-3852. Washington, D.C.: U.S. Environmental Protection Agency.

Wassermann, M., L. Tomatis, and D. Wassermann (1974) Storage maps of organochlorine compounds (OCC) in humans. In Epidemiological Deductions for Further Monitoring. CEC-EPA-WHO International Symposium-Environmental and Health, Paris 1974, Paper #115.

Webb, R.G. and A.L. McCall (1973) Quantitative PCB standards for electron capture gas chromatography. J. Chromatogr. Sci. 11:366-373.

Weiniger, D. (1978) Accumulation of PCBs by Lake Trout in Lake Michigan. Thesis. Madison, Wis.: University of Wisconsin.

White, D.H. (1976) Nationwide residues of organochlorine in starlings, 1974. Pestic. Monit. J. 10:10.

Whitmore, F.C. (1975) A Study of Pesticide Disposal in a Sewage Sludge Incinerator. Office of Solid Waste Management Programs, Final Report, Contract No. 68-01-1587, NTIS PB-253 485. Washington, D.C.: U.S. Environmental Protection Agency.

Williams, P.M. and K.J. Robertson (1975) Chlorinated hydrocarbons in sea-surface films and sub-surface waters at nearshore stations in the North Central Pacific Gyre. Fish. Res. Bull. 73:445-447.

Young, D.R. (1975) Inputs and Distributions of Chlorinated Hydrocarbons in Three Southern California Harbors. Report TM 214. El Segundo, Calif.: Southern California Coastal Water Research Project.

Young, D.R., D. McDermott-Erhlich, and T.C. Heesen (1975) PCB Limits to the Southern California Bight. El Segundo, Calif.: Southern California Coastal Water Research Project.

Young, D.R., D. McDermott-Ehrlich, and T.C. Heesen (1977) Sediments as source of DDT and PCB. Mar. Pollut. Bull. 8:254-257.

2

POLICY ASSESSMENT

The data presented in Chapter 1 indicate the extensive environmental distribution of PCBs and the existence of several localized hot spots for which removal or cleanup action may be justified. State and federal officials have been coordinating efforts to reduce PCB levels at the hot spots and to prevent future contamination (U.S. EPA 1978a,b; Hetling et al. 1978). This chapter will attempt to determine the costs of alternative control and cleanup policies and the potential benefits to human health and the environment.

CONCEPTS OF BENEFIT-COST ANALYSIS

Benefit-cost analysis evaluates beneficial and adverse effects of a proposed action in terms of a common monetary unit, like dollars. (For detailed discussions of benefit-cost analysis and its relationship to the underlying principles of welfare evaluation, see Prest and Turvey 1965, and Haveman and Weisbrod 1975.) Alternative actions are ranked according to net benefits, that is, benefits minus costs. Social good is defined in terms of the monetary value of net beneficial effects, without regard to whom the benefits accrue. The bulk of the practice of benefit-cost analysis involves the development and implementation of rules for measuring or inferring monetary values for beneficial and adverse effects where these are not directly observable.

Economic Benefits

Monetary benefits are a measure of changes in utilities or welfare of individuals. It has been said that to economists, "Man is the measure of all (economic) things"; thus, human preferences provide the basis for defining and measuring economic benefits. Only those things that individuals use either directly or indirectly can be included in a measure of benefits. This means that some effects cannot be measured in economic terms, e.g., certain ecological effects, such as alterations in the size of species populations, or subtle changes that affect freshwater or marine trophic levels. However, where such ecological changes affect commercially exploitable resources, such as fisheries or the use of ecological systems for outdoor recreation, the effects can be expressed in economic measures of benefits.

The benefits of controlling chemicals such as PCBs can accrue to individuals and society either directly in the form of increases in the availability of goods and services (e.g., improved health or improved recreation opportunities), or indirectly through increases in production efficiency. Assigning monetary values to benefits resulting from chemical control policies involves three distinct stages, each of which must be understood if the benefits are to be measured.

Stage One Reduction in the quantities of chemicals released into the environment leads to reduction in the concentration found in various environmental reservoirs. Modeling the functional relationships between changes in chemical releases and changes in ambient chemical levels is a complex matter, yet these processes must be understood before the benefits of chemical regulation can be estimated.

Stage Two Changes in ambient chemical concentrations lead to changes in human use of the environment, that is, to changes in the level and composition of services yielded by the environment. Understanding the relationships between changes in ambient quality and effects on human activities is an essential part of the estimation process and should be termed the measurement process to distinguish it from the process of valuation discussed in the next stage.

Stage Three The effects of changes in ambient quality have their counterpart in changes in an aggregate willingness to pay for uses of the environment. Here the economic concepts of demand and value are central to the analysis. However, the successful implementation of the valuation process of Stage 3 must often await the development of a better knowledge of the physical, ecological, and other processes governing Stages 1 and 2.

Costs

The costs of a regulatory action are defined and measured by determining which utility or benefit opportunities are foreclosed by the action. The control of chemical contamination involves the use of scarce resources that might otherwise have been put into different activities. Costs associated with these "new" uses are called "opportunity costs" and can be measured in monetary terms by society's or the aggregate willingness to pay for those items that must be given up.

How can the opportunity cost of regulation be measured? When regulation requires producers to use real resources such as labor, capital, and materials for the installation of treatment and control equipment, special handling of materials, or development of special disposal methods, the market values of these resources can be taken as measures of opportunity cost. For example, under certain conditions wages are a measure of the marginal productivity of labor in the economy and therefore reflect the value of other goods foregone when labor is diverted from one sector of the economy to efforts to control chemicals.

Regulatory actions may also reduce the availability of goods and services to consumers in a more direct fashion. For example, regulation may take the form of a ban on the use of a particular chemical or the production of particular goods using the chemical as an ingredient, or it may involve restrictions on allowable chemical tolerances in food (e.g., the FDA maximum PCB concentrations in fish). Such policies do not absorb resources directly, but they do impose costs in the form of losses in consumer welfare. In some cases, the cost of substituting alternative goods for those banned by regulation provides a basis for an estimate of true economic cost. A well developed economic methodology exists for making estimates of this sort.

Other Economic Impacts

Other types of economic impact associated with regulatory decisions can be described and quantified in monetary terms, but they are not commensurate with the economic efficiency benefits and costs described earlier. The incommensurability arises because the measures of impact are usually based on changes in the dollar volume of a particular class of economic transaction rather than on changes in the quantities of goods and services available to consumers. The distinction is similar to the one often made in the benefit-cost literature between real and pecuniary effects (Haveman and Weisbrod 1975).

To illustrate the point, consider the case of a chemical plant that closes because of regulatory action, thus terminating an annual million-dollar payroll. This surely has economic impact on a region, but it is not necessarily a real cost in the sense of a reduction in available goods and services. The relationship between the measure of impact and the measure of cost depends on the mobility of the labor force and other resources and on how speedily adjustments can be made. These factors will vary widely from case to case. Where a firm or an industry reduces output because of governmental regulations, there is a cost in economic efficiency only to the extent that, for a time, resources remain unemployed or underemployed. Lost output is an appropriate measure of cost in this case. The cost of resources absorbed in relocation or other activities designed to speed the adjustment process (e.g., labor retraining programs) should be taken into account. If all workers laid off from a plant find alternative employment immediately and at the same wage, there is no real cost other than the expense of relocation. If none of the labor force can find alternative employment, there is a cost in the form of lost output, and the lost payroll may be used as a measure of this cost.

COST-EFFECTIVENESS ANALYSIS

A cost-effectiveness analysis can be used in circumstances where a desired level of beneficial effect has been established and where the policy has been set to achieve that level at minimum cost. A cost-effectiveness criterion ranks alternative policies in terms of the cost required to achieve the given level. Because all the policies being evaluated achieve the same level, the beneficial effect will be the same and the various options will differ only in terms of their costs. One advantage of cost-effectiveness analysis is that problems created by the noncommensurability of benefits and costs are avoided. It should be noted, however, that whereas this type of analysis can be useful in evaluating alternative options for achieving a specific goal, it cannot be used to determine the choice of the goal itself.

EFFECTS ON HUMAN HEALTH AND THE ENVIRONMENT

As the data presented in Chapter 3 and Appendix D indicate, the major effects on human health and the environment resulting from exposure to PCBs take the form of subtle impairments rather than gross morphological or pathological changes. Acute exposure of either humans or wildlife rarely occurs, and those effects that have been observed are the result of cumulative contacts over long periods of time. The section entitled "Assessment of PCB Hazard" in Chapter 3 presents the scientific evidence relating to PCB toxicity and illustrates some adverse effects of unintentional release of the chemical to the environment.

Effects on Human Health

Exposure of humans to PCBs can occur through a number of different routes--occupational exposure, ingestion of contaminated foods or fish, exposure to contaminated soil or water, and transmission from mother to child through breast feeding. The levels of contamination are generally low, except where occupational exposure or direct contact occurs. Because of the low concentrations and the subtle nature of the toxic effect, it is extremely difficult to assess precisely those modifications that may have arisen in the health of human and animal populations as a result of environmental exposure to PCBs (see Table 3.6, Chapter 3). Well-documented cases of PCB intoxication in the human population have been limited to industrial accidents involving workers directly responsible for PCB production, excessive exposure to manufactured products containing PCBs, and consumption of edible products contaminated with PCBs during processing (Matthews et al. 1978).

Some of the nonspecific effects on health that may be attributed to low-level exposure to PCBs are abnormal fatigue, abdominal pain, numbness of limbs, swelling of joints, chronic cough, menstrual irregularity, and headache. Abnormal tooth development, hyperpigmentation, and low weight in newborn children also may be complications resulting from PCB exposure. Abnormalities in blood lipids, anemia, lymphocytosis, and adrenocortical hypofunction have been recorded in a number of chronic diseases associated with PCB intoxication. In addition to dermatological abnormalities, such as acne and hyperpigmentation, there have been suggestions of increased incidence of cancer in some of the Japanese who were exposed to PCB through contamination of cooking oil (Urabe 1977, Umeda et al. 1978).

Other potential nonspecific abnormalities that may occur as a result of chronic exposure to PCBs have been suggested by controlled experiments using nonhuman primates. In addition to alterations in the menstrual cycle and births of abnormally small infants, nonhuman primates experienced a greater frequency of early abortions following low-level exposure to PCBs (see Table 3.6, Chapter 3). Infants born to mothers exposed to PCBs during gestation and lactation also show some loss of immunological competence as well as learning and behavioral deficiencies. These abnormalities persist indefinitely.

Even though a large percentage of the human population has been exposed to varying amounts of PCBs, primarily through the food chain (i.e., contaminated fish, livestock, vegetation, and human milk), no well-documented cases of health problems associated with such environmental exposures exist. Thus, a detailed evaluation of the effects on human health in the general population is not currently available, although there are indications that even limited exposure to PCBs may produce injurious effects that will be difficult to detect even with close scrutiny.

Effects on the Environment

Assessment of effects on the environment is even less clear-cut than analyses of problems associated with health. The data in Chapter 1 suggest that PCBs are found in many places. However, as the results in Chapter 3 indicate, the toxic effects are subtle and difficult to detect. Reduced reproductive capability is the most common result of exposure to PCBs among many species. But other morphological and functional changes are noted as well, particularly in the livers of test animals. PCBs accumulate in adipose tissue, and it has been suggested that severe effects could arise when the animal is under sufficient stress to mobilize the PCB-containing lipids. Because PCB concentrations are magnified as the compound moves through food chains (see Appendix D, Bioaccumulation), the most pronounced toxic effects are noted in the upper trophic levels. Excessive accumulations may lead to reduction in the size of populations and to consequent alterations in the composition of communities and, ultimately, of the ecosystem.

ECONOMIC ANALYSIS OF PCB CONTROL STRATEGIES

In principle there are three categories of options for controlling or reducing potentially adverse effects of PCBs on health and the environment. They are:

- reducing the flow of additional PCBs into the environment (reducing or eliminating direct discharge of PCBs as an industrial waste product or controlling the use and disposal of materials currently in service that are contaminated with PCBs);
- controlling the exposure of populations (regulatory measures, such as limiting consumption of

fish, discouraging the practice of breast feeding, or setting strict limits on permissible levels of PCBs in foods); and

• reducing stocks of PCBs in environmental reservoirs (dredging river sediment, cleaning or upgrading existing landfills previously used for disposal of products containing PCBs, or enhancing biodegradation rates).

It is possible to assess, if only approximately, the cost and effectiveness of specific control strategies drawn from one or all of these three categories.

Cost Effectiveness of Control Strategies

The following specific control options have been considered for this report.

1. Governmental Regulations on Disposal of PCBs. EPA has published final regulations on the disposal of substances containing PCBs (U.S. EPA 1978a) and is currently considering revisions and additions to the regulations (U.S. EPA 1978b).

2. Dredging and Disposal of Contaminated River Sediment. The Department of Environmental Conservation in New York State has evaluated proposals to dredge sediment contaminated with PCBs from the upper Hudson River as a means of controlling further flow of PCBs into the estuary. Analyses are underway to determine the technological feasibility of effectively reducing PCB levels in various localized "hot spots" (Hetling et al. 1978).

3. Limitations of Permissible Residues in FDA Regulated Products. The FDA has implemented a temporary set of tolerances for PCB residues in several classes of food and currently has under consideration a proposal for further reductions (U.S. FDA 1977b). These limits apply to milk and dairy products (current limitation--2.5 mg/kg, proposed level--1.5 mg/kg fat basis), poultry (current limitation--5.0 mg/kg, proposed level--3.0 mg/kg fat basis), eggs (current limitation--0.5 mg/kg, proposed level--0.3 mg/kg), and fish and shellfish (current limitation--5.0 mg/kg, proposed level--2.0 mg/kg). The FDA reports that in all categories, except fish and shellfish, the observed contamination levels are well below the proposed limits. Therefore, the only

relevant portion of the proposed regulation would be associated with commercial distribution of fish and shellfish.

For each option we assess effectiveness in terms of the quantities of PCBs removed from the environment and destroyed or immobilized, or as in the case of the FDA regulations, the quantity of PCBs prevented from entering the human diet. In each case, we assume perfect compliance with the regulations. We also provide estimates of the cost-per-kg affected for each option. These estimates do not include administrative costs or the costs of inspection, policing and other enforcement activities.

Controlling Use and Disposal Table 2.1 summarizes the available information on costs of existing and proposed regulations governing use and disposal of PCBs. The values were derived from two reports commissioned by EPA (Versar, Inc. 1977, Westin et al. 1978), and from additional information provided by the Agency's Office of Toxic Substances. The data upon which the costs are based are presented in Appendix B.

The first column of Table 2.1 presents the quantities of PCBs that would be prevented from reaching the environment, either through destruction or immobilization, for each policy option. The second column gives the total cost of each policy option. The third column provides the average cost of each option in dollars per kilogram of PCB controlled. If within each option there were no variation in the cost-per-kg removed under different circumstances, then this figure could be taken as representing the incremental or marginal cost of removal for that option. Table 2.1 does not address all the EPA regulations, because some data on costs are not available. Also the proposed EPA regulations would impose no special requirements on disposal of small appliance capacitors or high intensity and fluorescent lighting ballasts, although estimates indicate that nearly 81.8 x 10^6 kg of PCBs are contained in these items and are currently in service (Westin et al. 1978). The Committee does not at present have estimates of the cost of controls on these PCB materials.

The ranking of options in Table 2.1 is based on the assumption that PCB releases prevented by each regulation would have resulted in equivalent

TABLE 2.1 Costs and Quantities Controlled for EPA Regulations on Use and Disposal

Policy	Quantity Controlled (kg)	Control Costs[b]	Average Cost Per kg
A. Incinerate fluids from PCB transformers (required)	123,000,000	$ 75,000,000	$ .61
B. Shred and incinerate high voltage capacitors (required)	90,900,000	$350,000,000	$ 3.85
C. Shred and incinerate low voltage capacitors (require)	31,800,000	$143,000,000	$ 4.49
D. Flush and drain transformers after A. (required)	12,300,000	$105,000,000	$ 8.50
E. Place transformers in chemical waste landfill (required)[a]	1,360,000	$ 45,100,000	$ 33.16
F. Replace PCB-containing electromagnets and incinerate fluid (proposed)	100,000-150,000	$ 2,800,000-3,400,000	$ 18.67-34.00
G. Replace PCB-containing			

motors in mining equipment (proposed)			
(1) Loaders	13,000	\$ 2,000,000	\$ 155.00
(2) Continuous Miners	1,136	\$ 640,000-2,240,000	\$ 563.00-1,972
H. Drain and incinerate mineral oil transformer fluids if contaminated by more than 50 ppm PCBs (proposed)[c]	10,900	\$612,000,000-769,000,000	\$70,000.00-88,000
I. Flush and replace contaminated hydraulic fluids in die casting (proposed)	data incomplete[d]	data incomplete[d]	\$133,000

[a]It is assumed that transformers will be disposed of in chemical waste landfills rather than by incineration.

[b]Options A through E will be undertaken as PCB-containing equipment is taken out of service. For example some PCB-containing transformers may remain in service for up to forty years. The control costs here will be incurred over the useful life of the equipment in question. These costs have not been discounted to the present. Discounting would reduce computed control cost, and would have the greatest effect on control options affecting transformers and transformer fluids. Discounting would have relatively little effect on options F through I, since these options would have to be undertaken shortly after the effective date of the proposed regulations.

[c]Fluids with greater than 500 ppm are covered by Policy A.

[d]Incremental costs are estimated from average costs per die-casting machine. Aggregate costs and quantity controlled are not known because of lack of data on the number of affected machines.

SOURCE: Modified from Versar (1977), Westin et al. (1978).

environmental and health risks. However, some forms of release could be potentially more threatening than others, especially to health. For example, continued use of PCBs in mining equipment (option G) could subject operators to significant risk of severe contamination. Even with this qualification, the data of Table 2.1 indicate a classic pattern of increasing marginal costs of control, rising to extremely high levels for the last increments of material controlled. The options listed would, if implemented, control a total of 260 x 10^6 kg of PCBs. Virtually all control is accomplished with options A through E at average costs of $33/kg or less. The remaining options listed (F through I) would contribute less than 0.1 percent to further control PCBs but would add significantly to the total cost; marginal costs of the last option rise to more than $100,000/kg of PCBs.

Dredging Hudson River Sediment The New York State Department of Environmental Conservation has examined several options for dredging contaminated sediment in the upper Hudson River and encapsulating it in special landfill sites. Information on these options is presented in Table 2.2. Maintenance dredging refers to routine maintenance of river channels. Although some PCBs would be removed by this dredging, the only portion of the cost that can be properly attributed to PCB control is that associated with encapsulation of the dredge spoils. During the summer of 1978, the state removed portions of the remnant deposits and stabilized the remainder. This appears as option B in Table 2.2. Each of the remaining policy options (C through E) involves more comprehensive and extensive dredging operations than option A, and thus greater costs.

Incremental costs and quantities of PCBs removed have been calculated from data presented by the New York State Department of Environmental Conservation (Hetling et al. 1978). These control costs are computed on the basis of encapsulation of the dredge spoils rather than destructive incineration of sediment contaminated with PCBs. If incineration were employed, costs would be higher. The data show increasing incremental costs as larger quantities of PCBs are removed. These dredging costs are higher than the least costly use and disposal regulations of EPA (options A through E, Table 2.1), but well below the most costly EPA proposed regulations. The cost-effectiveness measures for the Hudson River

TABLE 2.2 Costs and Quantities Controlled for Dredging Hudson River Sediments

Policy	Incremental Quantity Controlled (kg)[a]	Incremental Control Costs	Incremental Cost Per kg
A. Maintenance dredging	23,100	$ 2,500,000	$ 110
B. Partial removal of remnants and stabilization of remainder (performed during summer of 1978)[b]	7,700	$ 500,000	$ 65
C. Remove remaining stabilized remnant deposits	15,100	$ 3,300,000	$ 219
D. Hot spot dredging	77,000	$ 22,400,000	$ 291
E. Remove all river sediments	55,600	$175,300,000	$3,150
Cumulative Total	178,500	$204,000,000	

[a]Quantity controlled is based on revised data received from Toffelmire and Quinn, New York State Department of Conservation, personal communication, 1978.
[b]Remnant refers to sediment exposed by removal of the Fort Edward Dam in the summer of 1973.

SOURCE: Modified from New York State Department of Environmental Conservation (1978).

dredging options are based on the assumption that dredging and disposal of spoils do not of themselves pose significant environmental hazards (Hetling et al. 1978).

<u>Limiting PCB Content of Fish for Human Consumption</u> If a limit on the amount of PCBs in fish were imposed, the economic productivity of fisheries would be reduced. Fish with PCB content exceeding the limit would have to be destroyed, and it is likely that fishing for critical species in contaminated areas would cease.

The cost of imposing a maximum limit on allowable PCBs in fish would be the value of the reduction in the economic productivity of fisheries. Precise measurement of this loss is difficult in practice. For simplicity, it can be assumed that the effect of the limitation on the total output of fisheries would be so small that there would be no change in the price of fish. The FDA (U.S. FDA 1977a) economic analysis is based on this assumption, which seems plausible. At one extreme, the assumption would be that fishing activity continues at the same level, but that catches contaminated by PCBs in excess of the limit would be confiscated and destroyed. The resulting reduction in sales revenue at the dock (landed value) can be taken as a measure of the economic cost of the limitation. It has been estimated that the proposed 2 mg/kg temporary limit for fish would result in a loss in landed value of $1 million for marine fisheries and $7 million for freshwater fisheries, the latter located primarily in the Great Lakes region (U.S. FDA 1977a). However, landed values are likely to overestimate real net losses because at least some of the capital and labor, whose productivity is impaired, would be deployed elsewhere.

The assumption at the other extreme would be that particular fishing activities are decreased, but that capital and labor in this fishery would be able to move without productivity losses to other activities, such as fishing for other species or moving to other waters. In this case, the loss in output of fisheries would be exactly offset by the increase in output of other areas as resources would be diverted from one use to another.

Neither extreme is likely to be an accurate representation of the real world. There will be changes in levels of activity and movements of resources in response to any limitation on PCBs. But capital and labor in fisheries are specialized resources and not perfectly mobile. Shifts in resources will occur, but only with a time lag and some reduction in productivity. Precise estimates of cost associated with limiting PCB concentrations in fish would require better data and more careful analysis than we are able to provide. Therefore, for purposes of this study, $8 million is taken as an upper-bound estimate of the economic cost to commercial fisheries resulting from the FDA limit on PCB levels in fish.

The economic impact can be translated into a crude estimate of the cost-per-kg of PCBs avoided in the human diet as a result of the regulation. If, as a first approximation, it is assumed that the average landed price of PCB-contaminated fish is $0.44/kg, an $8 million loss in landed value is equivalent to a reduction of 18.2×10^6 kg in the catch. Since the FDA proposal would lower the allowable PCB content from 5 mg/kg to 2 mg/kg, we assume that the average concentration in affected fish would be 3.5 mg/kg. Thus, the limitation would reduce total PCB input to the

American diet by 64 kg. Thus $125,000 would be the upper-bound estimate of the cost-per-kg eliminated from edible fish. FDA also estimated that the incremental loss in landed value from lowering the tolerance from 2 mg/kg to 1 mg/kg would be $10 million. By similar calculations, the cost for avoiding an additional 34 kg of PCBs in the human diet would be $293,000/kg.

The FDA limitation applies only to fish entering interstate commerce. If individual states also imposed effective limitations on PCB content of fish caught and consumed by sportsmen, the total cost of controlling the intake of PCBs from fish would be higher but not necessarily the cost-per-kg.

The cost of regulating sports fisheries could take two forms. If the regulation applied only to consumption of fish and not to the fishing activity itself, the utility and value of time spent fishing would be reduced. Even with no change in total sports fishing activity, this reduction in value would constitute a regulatory cost. To the extent that recreational fishing activities were reduced, there would be a further loss equal to their value per day multiplied by the number of days of lost activity. Quantitative estimates of the magnitude of these costs are not available. Total spending on sports fishing activities (for such items as food and lodging during trips, gear, boats, or bait) is not a measure of the potential economic cost of a sports fishing ban, because reduced spending on fishing probably would be offset by increases in other categories of expenditure. Productive resources would be reallocated, and if resources were perfectly mobile, there would be no net loss in the output of the national economy in goods and services.

Cost-Effectiveness of Alternatives

The preceding sections provide estimates of cost-effectiveness for three different categories of control policy, using different measures of effectiveness and allowing for a ranking of options within each control category. The cost-effectiveness measures would be more useful, however, if they could be made commensurable across categories. If a kilogram of PCBs released through unregulated disposal were equivalent in environmental impact to a kilogram of chemical present in river sediment, the cost effectiveness of measures for destroying PCBs and for dredging would be commensurable. If that assumption were to hold, the first six EPA regulations (options A through F) in Table 2.1 would be more cost effective than any of the dredging options. But if EPA were to approve all the proposed regulations (options F though I, Table 2.1), including those dealing with contaminated transformers and die-casting fluids, removal of all Hudson River sediment also would be economically justified. Dredging activities would produce equivalent results in terms of PCB control at costs considerably below those expected of the most expensive EPA regulation.

However, the control of PCBs through dredging options may not be fully equivalent to regulating PCB use and disposal. The New York State Department of Conservation estimates that about one-third of the mass of PCBs in sediment in the Upper Hudson is subject to erosion by river currents (Hetling et al. 1978). Given this estimate and assuming that dredging could not be selectively applied to the most mobile sediment containing PCBs, it would be necessary to remove perhaps 3 kg of sediment for an effect equivalent to destroying 1 kg of PCBs presently in service. To make the costs of dredging comparable to values in Table 2.1, incremental measures of cost for complete dredging would be increased by perhaps a factor of 3. Removal of remnants and hot spot dredging are more selective operations, and measures of cost effectiveness for these options are probably closely comparable to those for regulation of use and disposal.

To make the measures of cost-effectiveness for dredging and regulations of disposal commensurable with the food tolerance proposals, it is necessary to estimate what fraction of PCBs released to the environment would eventually accumulate in fish and shellfish destined for human consumption. The following calculation was used to estimate this fraction. In the Great Lakes, the total quantity of PCBs in sediment is on the order of 3.5×10^6 kg (Table 1.16; St. Lawrence drainage, upper bound estimate). Fishing removes about 2.7×10^2 kg/yr (about 9.0×10^7 kg contaminated at typical PCB levels of about 3 mg/kg). Assuming that such levels will be maintained for 30 years before PCBs are buried or dissipated, about 8×10^3 kg of PCBs would be incorporated in fish and consumed by humans. Hence, eliminating 1 kg of PCBs from the human diet would require the removal of about 4 x 10^2 kg from the environment. The figure could be reduced if the sediment deposits most likely to contribute to residues in fish could be removed selectively, as discussed above for the Hudson River.

On the basis of these admittedly rough calculations, a conservative estimate is made that 0.25 percent of all PCBs released to the environment enter the diet. This figure was derived from data on the Great Lakes, a relatively closed system; similar calculations for the Hudson River biota indicates roughly 0.30 percent of PCBs released into the environment might enter the diet (Hetling et al. 1978). It may be expected that in more open areas, such as estuaries and oceans, a smaller fraction of PCBs would enter the human diet. Thus, the measures of cost effectiveness for the options of regulating PCB use and dredging can be made commensurate with the FDA food tolerance proposals by raising them by 400 or more. In other words, removing 1 kg from the human diet at a cost of $125,000/kg is equivalent to removing 400 kg or more from the environment at a cost of $312/kg. (It should be noted that this adjustment implies that there are no beneficial effects of dredging and regulation other than impacts on the human diet.)

On the basis of these computations, either of the FDA food tolerances (2 mg/kg or 1 mg/kg) is more cost effective in protecting humans against exposure to PCBs in the diet than the more costly of the EPA proposed regulations for use and disposal. However, hot-spot dredging may be a preferred option both because of slightly lower estimated costs-per-kg removed and because of the possibility of preventing other damages in addition to those to human health. This is not to say that the FDA regulations should be approved or that incinerating contaminated transformer fluids is too costly for policy consideration. Cost-effectiveness data do not constitute an adequate basis for making policy decisions about control options. Although a ranking of options by effectiveness per dollar of resource committed is possible, these data do not tell us how far out and up along the incremental cost curve we should go. To make such determinations, further information is needed on the beneficial effects of controlling PCBs.

<u>Benefits Versus Costs</u> It is desirable to estimate the expected benefits to allow for comparison with the costs

of reducing releases of PCBs (Table 2.1) or of reclaiming PCBs already released (Table 2.2). Major categories of benefits derived from control options include: reduced risks to human health, enhanced economic benefits to commerical and recreational fishing activities, and reduction in pollution damage to the environment. We are not able to make any quantitative assessments of the last two categories, and even a tentative estimate of benefits to human health is incomplete and subject to uncertainty.

Many of the risks to human health resulting from PCB exposure are perceived as subtle physical and behavioral changes. Recently it has also been suggested that exposure to PCBs poses a threat of cancer to humans (Umeda et al. 1978, Urabe 1977). Numerical estimates are available for the latter type of risk only.

The National Academy of Sciences' Safe Drinking Water Committee (National Research Council 1977) estimated that the lifetime carcinogenic risk to a person ingesting p micrograms of PCBs daily (in drinking water) would be approximately 3×10^{-6} p. The estimate was derived by linear extrapolation of results from a bioassay for carcinogenesis using rats exposed to Aroclor 1260 at 100 mg/kg in the diet. This estimate has a high degree of uncertainty, for three reasons.

1. The estimate was derived by extrapolating over an unusually wide range of exposures (3 orders of magnitude even for highly exposed persons);
2. No scaling factors were incorporated for extrapolation from potential risks of cancer for rats exposed to low doses to expected risks of cancer for humans exposed to corresponding levels of PCBs; and
3. The estimate was derived only for pure grade Aroclor 1260, without detectable impurities; environmental residues of technical grade PCBs may be more or less hazardous. (See the discussion of PCB impurities in Appendix D.)

The daily ingestion of p micrograms of PCBs would result in a lifetime intake of about 2.5×10^4 p micrograms. Using the above relationship between exposure and risk, the incremental cancer risk per microgram of PCBs is approximately 1.2×10^{-10}. If the exposure-risk relationship is assumed to be linear (at least at the margin), the estimate of individual risk may be extended to a population. Neither the distribution of exposures within the population nor the distribution with age need be considered. Hence, on an average, one additional cancer case would be expected to result from an incremental intake of 8 x 10^9 μg of PCBs; however this intake may be distributed among the U.S. population.

We assume that the economic benefits to society of reducing a general risk of cancer range from $100,000 to $1,000,000 per cancer death avoided. (The basis for this is discussed in Appendix C.) Given this assumption, the benefits assigned to the reduction of PCB intake are in the range of $12/g to $120/g ($12 x 10^3/kg to $120 x 10^3/kg).

To reduce human intake by 1 g, a larger quantity of PCBs must be removed from the environment or withheld from discharge. As discussed above, roughly 400 g or more of PCBs must be withheld to reduce dietary intake by 1 g. Thus, the marginal benefits to be gained from removing PCBs from the environment would be in the range of $30/kg to $300/kg.

In addition to the major qualifications stated above, several uncertainties about this estimate should be emphasized.

- The estimate refers only to carcinogenic risks, although other types of risks to health and other forms of benefits may be more important (see Hazard Evaluation, Appendix D for PCB toxicity).
- The costs of PCB removal are compared with putative benefits that might not be realized for 30 to 80 years. If it is considered appropriate to discount benefits over such lengthy periods (2 to 3 generations), the amount discounted might be in the order of a factor of 100 or more.
- The assumption is made that PCBs released to the environment now will continue to accumulate in fish for approximately 30 years (potential residence time of PCBs in sediment; see Chapter 1). If this estimate is too long, benefits resulting from control options would be correspondingly overestimated. If the estimate is too short, the calculated benefits would be determined primarily by the discount rate and not by persistence of the PCBs.

For these reasons, the benefits of controlling PCBs might be overestimated or underestimated by a factor of 100 or more. Accordingly, although the estimates are given as \$30/kg to \$300/kg for benefits derived from removal of PCBs, it is not possible to state with confidence that these costs are not much higher or much lower.

Considering the great uncertainties in these estimates, health benefits of \$30/kg to \$300/kg of PCBs removed would economically justify only the least costly EPA disposal regulations on strict benefit-cost grounds (Table 2.1). The FDA proposed food tolerance limit may also be justified at this benefit level because the cost involved, adjusted for the ratio of environmental to dietary levels, is only \$312/kg. If we recall that the benefit estimate is only for one category of health effects and that it may underestimate expected benefits by two orders of magnitude, economic justification for incurring incremental costs of \$30,000/kg may be possible on the basis of potential benefits. Several of the EPA proposed regulations and the selective Hudson River dredging options also may be justified since their costs lie within the estimated range of benefits. However, given the range of uncertainty, we cannot rule out the possibility that costs may exceed benefits for all control options considered.

When regulatory decisions such as those assessed here must be made in the face of substantial uncertainty, the final choices involve value judgments and implicit trade-offs by the responsible public officials (see National Research Council 1975 for a more comprehensive discussion of this point). But one additional set of calculations can be made in an effort to illustrate some of the value implications of alternative choices.

In 1968 more than 1,000 people in Japan were exposed to high dietary levels of PCBs from the accidental contamination of rice oil in what has become known as the Yusho incident. The average cumulative dose was estimated to be 1 g/person. (See Chapter 3 for a description of this incident.) The cost of preventing 1 kg of dietary intake (sufficient to provide a Yusho dose to 1,000 people) by lowering PCB limits in fish would be \$125,000, or \$125 per person.

If this were judged as a reasonable price for preventing the magnitude of effects observed in the Yusho victims, the FDA policy as well as maintenance dredging and the other lower cost EPA control options could be economically justified. But by this line of reasoning, a justification for the highest cost EPA disposal regulations would require a willingness to pay $53,200 per person to avoid PCB exposures similar to the Yusho dose. However, three qualifications should be made.

1. The PCB levels observed in fish would not provide the Yusho dose to small numbers of people at currently typical levels of fish consumption. Rather larger numbers of people would receive only 1 to 2 percent of the Yusho dose. Thus, the above comparison would be valid only if the dose-damage relationship were linear throughout this range.

2. The Kanechlor involved in the Yusho incident was contaminated with unusually high levels of polychlorodibenzofurans; these may have been responsible for some portion of the observed effects on health (Miyata et al. 1977).

3. Accumulation of the 1 g dose by the Yusho victims occurred over a shorter period than would be typical for a person whose diet contained fish contaminated with PCBs.

Tables 2.1 and 2.2 indicate high average or incremental costs with some options for recovering PCBs already discharged or capturing those currently in use to prevent their eventual discharge to the environment. This raises the question of whether it might have been less costly for society to have avoided these problems by preventing the initial production and use of PCBs.

The cost of such an initial prospective ban is measured by the additional cost of achieving equivalent product performance through redesigning equipment and substituting other materials. For those examples where redesign and substitutions are not feasible, we must assess cost factors related to reduced product performance or the maintenance of safety. We cannot accurately estimate these costs, but we can provide an illustrative calculation.

The range of the current cost estimate for banning the use of PCBs in transformers and capacitors is $14.7 million to $32.6 million per year (Westin et al. 1978). This represents primarily the costs of substituting materials, redesigning products, and increasing risks of fire. Using the upper limit, we assume for purposes of illustration that the total cost for all segments of the economy using PCBs is 4 times this amount. These are annual costs. It is interesting to note the discounted present value of a time stream of costs evaluated, hypothetically, at the time of the first introduction of PCBs into commercial use. If we assume infinite time and a discount rate of 8 percent, the current value of costs to ban PCBs would have been \$1.6 billion. Cumulative production of PCBs has amounted to about 650 x 10^6 kg since 1930 (Westin et al. 1978). Thus the cost per kilogram of avoiding that production and its subsequent damage to the environment would have been about \$2/kg. This is at the low end of the range of cost-effectiveness measures.

The calculation suggests that prevention can be more cost effective than control and cleaning up measures. But this should not be interpreted as justifying a decision not to have allowed PCB production to begin for two reasons. First, the costs of such a ban would have to be compared with its anticipated benefits. Second, it is not appropriate to think of the choices as being limited to the two alternatives of no production versus production and use in the current patterns. Instead one should attempt to perform separate benefit-cost calculations for each type of use. Use of PCBs in closed systems with strict controls to prevent distribution of PCBs in the environment might be feasible in benefit-cost terms, especially for those uses where PCBs are particularly valuable. As Table 2.1 shows, costs of collection and destruction for closed system uses are relatively low. The high control costs are associated with recovery of those PCBs that have contaminated other materials or escaped to the environment.

CONCLUSIONS

1. Substantial differences exist in the costs of various options for the control of PCBs. Among the alternatives analyzed--which included EPA proposed regulations on disposal and use, options for dredging and disposal of contaminated river sediment, and FDA regulations limiting permissible residues in fish destined for human consumption--average costs varied by as much as 5 orders of magnitude.

2. Existing EPA regulations on disposal and use are projected to result in the control of 260 x 10^6 kg of PCBs with average costs of up to \$33/kg of PCBs controlled. Proposed EPA regulations would increase the

total quantity of PCBs controlled by about 0.1 percent (beyond those controlled by existing regulations) with average costs becoming as high as $133,000/kg.

3. A case study of the Hudson River indicated that various alternatives for dredging sediment could remove a total of about 178 x 10^3 kg of PCBs from potentially mobile river sediment at costs ranging from $110/kg to $3,150/kg.

4. It is estimated that up to 0.25 percent of the total PCBs released to the environment enter the human diet; thus at least 400 kg of PCBs would have to be removed from the mobile environmental reservoir in order to achieve a reduction of 1 kg of PCBs in food. This figure (400 kg) is derived using data from the Great Lakes, a closed system; it may be too low for more open systems. FDA limitations on PCB concentrations in commercially distributed fish could reduce PCB input to the American diet by about 64 kg at a cost of less than $125,000/kg. (A reduction of an additional 34 kg would cost less than $293,000/kg.)

5. Based on the figures used in the analysis in this report, FDA regulation of PCB limits in fish destined for human consumption appears a more cost-effective method of controlling dietary intake and human health effects than some of the most costly EPA proposed regulations on use and disposal. Some dredging options may be preferable to either EPA or FDA regulations.

6. It is difficult to assess the benefits of controlling exposure to PCBs with currently available information. The analysis in this report estimates the benefits as $30/kg to $300/kg of PCBs removed from the environment or withheld from discharge, but notes that even these figures could be high or low by a factor of 100 or more.

RECOMMENDATIONS

1. Policy makers should use available data to calculate the incremental or average costs of PCB control options in conducting economic analyses of proposed regulations.

2. As greater control of PCBs is sought, regulatory efforts should be concentrated on those options which are most cost effective, that is those with the lowest average or incremental costs per comparable effect or benefit.

3. Economic analyses should include cost data on all options--not just those under a single agency's

regulatory authority. The analysis could then show--as does the analysis presented in this report--that some options beyond a particular agency's authority may be substantially more cost effective than some that lie within the agency's jurisdiction. A comprehensive, coordinated government policy toward control of pollutants should reflect such findings.

REFERENCES

Haveman, R.H. and B.A. Weisbrod (1975) The concept of benefits in cost-benefit analysis: With emphasis on water pollution control activities. Pages 37-66, Cost-Benefit Analysis and Water Pollution Policy. Edited by Henry M. Peskin and Eugene P. Seskin. Washington, D.C.: Urban Institute.

Hetling, L., E. Horn, and T.J. Tofflemire (1978) Summary of Hudson River PCB Study Results. Technical Paper No. 51. Albany, N.Y.: New York State Department of Environmental Conservation.

Matthews, H., G. Fries, A. Gardner, L. Garthoff, J. Goldstein, Y. Ku, and J. Moore (1978) Metabolism and biochemical toxicity of PCBs and PBBs. Environ. Health Perspect. 24:147-155.

Miyata, H., T. Kashimoto, and N. Kunita (1977) Detection and determination of polychlorodibenzofurans in normal human tissues and Kanemi rice oils caused 'Kanemi Yusho.' J. Food Hyg. Soc. 18(3):260-265.

National Research Council (1975) Decision Making for Regulating Chemicals in the Environment. Committee on Principles of Decision Making for Regulating Chemicals in the Environment, Environmental Studies Board, Commission on Natural Resources. Washington, D.C.: National Academy of Sciences.

National Research Council (1977) Drinking Water and Health. Committee on Safe Drinking Water, Assembly of Life Sciences. Washington, D.C.: National Academy of Sciences.

New York State Department of Environmental Conservation (1978) Toxic Substances Impacting on Fish and Wildlife. Division of Fish and Wildlife Monthly Report 12, March 20. Albany, N.Y.: New York State Department of Environmental Conservation.

Prest, A.R. and R. Turvey (1965) Cost-benefit analysis: A survey. Econ. J. 75(4):683-735.

Umeda, G., H. Nakajima, and K. Yatsoki (1978) Some problems on health effects of polychlorinated biphenyls. Ecotoxicol. Environ. Saf. 1:437-446.

Urabe, H. (1977) Foreword, the Fourth Report of the Study on Yusho and PCB. Fukuoka Acta Medica 65:1-4.

U.S. Environmental Protection Agency (1978a) Polychlorinated biphenyls (PCBs): Disposal and marking. Federal Register 43(34):7149-7164.

U.S. Environmental Protection Agency (1978b) Polychlorinated biphenyls (PCBs): Manufacturing, processing, distribution in commerce, and use bans. Federal Register 43(110):24802-24818.

U.S. Food and Drug Administration (1977a) Economic Impact Assessment for Proposed Reduction of Temporary Tolerances for Polychlorinated Biphenyls in Food. (Unpublished paper)

U.S. Food and Drug Administration (1977b) Polychlorinated biphenyls (PCBs): Unavoidable contaminants in food and food packaging materials; reduction of temporary tolerances. Federal Register 42(63):17487-17494.

Versar, Inc. (1977) Microeconomic Impacts of the Proposed Marking and Disposal Regulations for PCBs. NTIS Report No. PB-267 833/2WP. Springfield, Va.: National Technical Information Service.

Westin, R.A., L. Fourt, D. Berkey, and B. Woodcock (1978) Microeconomic Impacts of the Proposed "PCB Ban Regulation." National Technical Information Service, NTIS Report No. PB 281-881/3WP. Springfield, Va.: Versar, Inc.

3

GUIDELINES FOR EVALUATING THE POTENTIAL TOXICITY OF CHEMICALS

After passage of the Toxic Substances Control Act, a concerted effort was made by various governmental agencies and nongovernmental groups to develop testing guidelines. The guidelines describe the type of data required for an agency evaluation of the potential adverse effects resulting from the manufacture and use of new chemicals. The purpose of this chapter is to determine whether the data required by the guidelines are sufficient for and relevant to making an adequate assessment of potential hazards to human health and the environment.

Draft guidelines were examined that had been prepared by three different sources: the Office of Pesticide Programs (FIFRA, U.S. EPA 1978a), the Office of Toxic Substances (TSCA, U.S. EPA 1978b), and the Conservation Foundation (1978). Basic data requirements were consistent among these drafts; variations generally depended on the intended use of the chemical (e.g., pesticide versus commercial chemical). A summary of the governmental requirements was prepared, and PCB data were compiled accordingly (Appendix D).

The evaluation was performed as though PCBs were new materials. Thus, when the requirements for particular

TABLE 3.1 Proposed Uses of Polychlorinated Biphenyls, 1930

Protective coatings (Aroclor resins in lacquers)
Investment die-casting (to replace natural waxes)
Heat-transfer agents

SOURCE: Taken from Penning (1930).

data depended on anticipated exposure or use patterns (e.g., teratogenic, reproduction, or some chronic testing), decisions about whether to include these data in the assessment were made based upon information regarding the original uses of PCBs (Table 3.1), rather than upon our knowledge of the ultimate exposure and environmental distribution. By 1930 commercial PCBs were being considered for use in protective coatings, such as varnishes and lacquers, in investment die-castings, and as heat-transfer agents (Penning 1930). Testing was under way to determine suitability for other uses, such as waterproofing, textile flame-retardant, and artificial leather finishing, but full-scale expansion of PCB use did not occur until much later.

GENERAL TESTING REQUIREMENTS FOR EVALUATING TOXIC SUBSTANCES

A comparison of the proposed EPA testing guidelines was made to identify those elements common among drafts. Test protocols were similar for most required data. TSCA guidelines followed FIFRA requirements rather closely. The major differences between the guidelines involved the prescribed sequences of hazard evaluation tests and criteria for the choice of particular tests. Summaries of the guidelines data requirements are presented below followed by comments of the Committee wherever appropriate.

Chemical Identity

Product Identity and Disclosure of Ingredients Complete information on the components of the product is required. It should include, to the extent possible and relevant, ingredients of manufacture and final product compounds that can reasonably be identified. The required information includes:

1. the chemical name from the Chemical Abstracts 1972 to 1976 Index of Nomenclature, or other well-defined names;
2. the Chemical Abstracts (CAS) registry number;
3. the product name, trade name, and common name (if established);
4. the experimental or internal code number;
5. the molecular (empirical) formula and the weight or the molecular weight range for each ingredient; and
6. the structural formula for each organic ingredient.

> Committee's Comments: It is uncertain whether a firm order of presentation will be established for this information. However, because many substances submitted for approval will be mixtures such as PCBs, the following sequence may be most appropriate for identifying a product:
>
> 1. the product name, trade name, and common name;
> 2. the experimental or internal code number;
> 3. for each ingredient:
>
> • the chemical name from the *Chemical Abstracts Index of Nomenclature*;
> • the Chemical Abstracts (CAS) registry number;
> • the molecular formula and structural formula (for organic, organo-inorganic, and inorganic compounds whenever appropriate); and
> • the molecular weight or weight range.

Description of Manufacturing Process Guidelines require that application for approval of manufacture or new use, or both, include information regarding the basic manufacturing process as well as information on the composition of the starting and intermediate materials.

Formation of Unintentional Ingredients An indication of whether unintentional ingredients are formed is required along with information on intentionally added ingredients; identification of components that might reasonably be detected as present in amounts of 0.01 percent or more of product weight are to be included.

> Committee's Comments: A variety of chemical contaminants is present in manufactured products in amounts less than 0.01 percent of product weight

(100 μg/g). For example, as shown in Table 3.2, the amount of chlorodibenzofurans (CDFs) reported in individual samples of commercial PCB products is substantially below this limit, and low-level exposures to CDFs have been reported to produce toxic effects in animals (Zitko et al. 1973, Nisbet 1976, Oishi and Hiraga 1978, Oishi et al. 1978, International Agency for Research on Cancer/IARC, in press). Thus a listing of all identifiable "contaminants" or unintentionally added ingredients should be required to make possible a proper evaluation of the potential toxicity of new compounds. For those ingredients known to be or suspected of being hazardous, an expected maximum limit within the product should be stated, even if this limit is less than 100 μg/g of product weight.

Declaration and Certification of Ingredient Limits
Estimates of upper and lower limits should be established for each ingredient as well as for any impurities known to be present. The limits should be presented as percentages of the total product.

Committee's Comments: In addition, quantities should be established for impurities that may be expected to arise during normal use of the product and under specified disposal conditions.

TABLE 3.2 Amounts of Chlorodibenzofurans Detected in Commercial PCB Mixtures

		μg/g
Aroclor	1248	2.0
	1254	1.7
	1260	1.0
	1016	ND (less than 0.001)
Clophen	A-60	8.4
Phenoclor	DP-6	13.6
Kanechlor	300	1.0
	400	18.0
	500	4.0
	600	5.0

ND indicates not detected.

SOURCE: Modified from Nisbet (1976).

Analytical Methods General requirements include descriptions of analytical methods (or references to commonly recognized analytical techniques in the published scientific literature) for detecting and measuring the concentration of ingredients or impurities. The techniques should include extraction procedures, confirmatory methods, and validation procedures, including background (or blank) values and recovery values; and they should be capable of detecting and measuring ingredients or impurities in quantities of 0.01 percent or more of total product weight.

> Committee's Comments: In addition to methods for product analysis, methods for detecting and determining ingredient or impurity levels in environmental media should be required for subsequent use in monitoring programs. Detection and measuring capabilities must be in the range of toxicologically significant quantities, if they are known, rather than at a limit of 0.01 percent of product weight.

Physical and Chemical Data

Data are required on the physical and chemical properties of the technical grade of each active ingredient present in the product. These must include color, odor, melting point, solubility, stability, octanol/water partition coefficient, physical state, density or specific gravity, boiling point, vapor pressure, pH, storage stability, flammability, oxidizing or reducing action, explosiveness, miscibility, viscosity, corrosion characteristics, and dielectric breakdown voltage. The FIFRA guidelines apparently require data on all these properties for each new chemical. TSCA requirements are more flexible, depending on use of the chemical, testing condition, and extent of measurable parameters.

> Committee's Comments: Required data should be relevant to assessing the hazard potential of the substance. They should be presented for known impurities as well as active ingredients. In the context of assessment of the chemical's environmental fate, the reactions of interest are those that determine the stability of a compound

under environmental conditions (0-40°C, normal atmospheric pressure, pH 4 to 10, presence of water, oxygen, ozone, clays or other sediments and minerals, or sunlight).

The stability assessment may be qualitative, based on a comparison of physical and chemical properties of the substance with those of other chemicals whose environmental properties and effects are known, or the analysis may be quantitative, based on models and quantitative structure-activity relationships.

Mobility of the substance in the environment can be judged qualitatively from its physical properties, such as water solubility, boiling point or vapor pressure, or octanol/water partition coefficient. Quantitative estimates may be possible also, based on physical and chemical properties as well as additional data, such as the volatilization rate constant/re-aeration rate constant ratio, and partition coefficients for natural sediments (Smith et al. 1977).

A variety of reactions including oxidation, reduction, hydrolysis, elimination or substitution may affect the substance in the environment. The science of reactions and reactivities is still largely in the empirical stage, but it is generally possible to predict appropriate pathways by considering functional groups present in the molecule. When reaction rates have been measured for suitable model compounds, correlations of reactivity based on electronic and steric factors can be used to estimate reactivity values. Although physical-organic chemistry provides tools for making predictions, more data on model compounds are needed to increase the reliability of the predictions.

A more complex situation occurs with photochemical reactions. Here again, photochemists can generally predict from the functional groups present in a molecule and from the visible and ultraviolet spectrum (which is generally also predictable from the structures), whether the molecule will absorb actinic light and therefore be photoreactive. On the other hand, there are no predictive correlations regarding quantum yield; when several reaction paths are available to an excited molecule, there is little or no ability to make predictions regarding the relative importance of the various reaction channels. Whereas many photochemical reactions have quantum yields less than unity (and most are accompanied by considerable energy wastage, so that a large fraction of the light absorbed is simply degraded to thermal energy), there are some that produce free radicals and may thus involve light-induced chain reactions. Chain lengths in such reactions are complicated functions of initiation, propagation, and termination rates and therefore, are not predictable with any degree of confidence in uncontrolled environments. For these reasons it generally will be necessary to do experimental work on the assessment of persistence and the hazards posed by new chemicals in the environment.

Environmental Data

The proposed guidelines require a variety of data with which to evaluate the potential environmental fate of a new chemical. Both FIFRA and TSCA require the types of tests described below; differences between the proposed guidelines are noted wherever appropriate.

Chemical Degradation Degradation studies are required to determine the rate of decomposition and to identify chemical residues that may adversely affect nontarget organisms. Oxidation and hydrolysis tests noting pH- and temperature-dependent properties must be made in both soil and aquatic environments. The TSCA requirements specify that only acid- or base-catalyzed reactions need to be considered. Atmospheric oxidation tests are to be performed only on organic chemicals with vapor pressures greater than 10 mm Hg at 30°C. Oxidation tests in aqueous solutions are required for water-soluble chemicals (solubility $\geq$ 1 mg/l at 20°C).

Photolysis should indicate intensity and wavelength of radiation. Data must include a material balance, half-life estimate, and the identification of each photoproduct present in amounts greater than 10 percent of the product weight. TSCA exempts from photochemical degradation testing those chemicals that have extinction coefficients of less than 0.1 $M^{-1}cm^{-1}$ at wavelengths greater than 290 nm.

Committee's Comments: Degradation studies should be used to assess effects of the chemical on the total environment. Data on hydrolysis, oxidation, reduction, and photodegradation should be presented as applicable. Qualitative predictions of potential degradation and toxicity of new chemicals are possible through (1) identification of functional groups in the molecule, (2) knowledge of general degradation processes, and (3) toxic properties of similar chemicals. The confidence of such predictions will improve with a better data base. However, although they will prove valuable in determining priorities in the sequence of toxicity tests, the predictions cannot replace the toxicological tests themselves.

The fate of the chemical undergoing reaction also must be considered. The environmental problem does not end when the chemical itself is transformed, unless it is completely mineralized or converted to species available for normal biological utilization. Transformation products may be more persistent and more hazardous than the parent compound, and thus must be dealt with in any analysis.

Biodegradation Microbial metabolic transformation may be the major route of degradation in soils and in the aquatic environment. Transformation data must be obtained under aerobic and anaerobic conditions in both soil and water. Information regarding the chemical's effect on microorganism development (bacteria, algae, diatoms, etc.) is required. Data concerning the disruption of such activities as cellulose decomposition, nitrogen fixation, sulfur transformation, and CO_2 efflux are particularly important. In addition an assessment of potential disruption of wastewater treatment systems must be made.

Committee's Comments: Microbial degradation is probably the major chemical transformation route in soils and aquatic sediment. In addition to the required data, estimates should be made of the biodegradability index (biological oxygen demand [5 days]/chemical oxygen demand = BOD_5/COD) and compatibility with biological waste treatment (IC_{50} = 50 percent inhibition of activated sludge) under both aerobic and anaerobic conditions. Detailed studies of biodegradation intermediates should be required for compounds with a low biodegradability index or a low compatibility with waste treatment, or both.

Mobility Movement of a product through the environment may cause contamination of food webs, loss of usable land and water resources, and loss of wildlife habitats; data requirements include information on leaching, volatility, adsorption/desorption, and water-dispersal properties.

Bioaccumulation Data must indicate the extent of accumulation and potential adverse effects on nontarget organisms (in particular, agricultural systems and fish). Residue levels and rate of uptake in plants must be determined according to the TSCA guidelines. Fish bioaccumulation studies should use flow-through and static systems with sunfish and catfish as the FIFRA recommended test species. TSCA requirements specify that fish bioaccumulation should be assessed when octanol/water partition coefficients are greater than 1,000 or when solubility in water is less than 0.9×10^{-3} M at 20°C and degradation data indicate persistence. Minnows and sunfish are preferred species for the TSCA guidelines.

Committee's Comments: At least two accounts of relationships between octanol/water partition coefficient and solubility in water have been published. The relationships derived by Chiu et al. (1977) and Moriguchi et al. (1976) give solubilities of 0.966×10^{-3} M and 0.621×10^{-3} M, respectively, for a partition coefficient of 1,000. The solubilities correspond to an accumulation coefficient of 80 to 100 (Chiu et al. 1977). The guidelines could be modified to require bioaccumulation tests with fish only for those compounds with a partition coefficient of 10,000, corresponding to a solubility of 0.02×10^{-3} M to 0.03×10^{-3} M, and an accumulation coefficient of about 500.

TABLE 3.3 Proposed Testing Levels for Hazard Assessment

Level I Tests	Level II Tests	Level III Tests
A. Acute Toxicity	A. Chronic Toxicity	A. Environmental Simulation
B. Short-term Screening	B. Reproductive Tests	B. Metabolic Tests
C. Environmental Chemistry	C. Bioaccumulation	C. Long-term Chronic Toxicological Tests

Hazard Assessment

The FIFRA proposed guidelines of June 1978 follow a hierarchical testing scheme for hazard assessment. Table 3.3 summarizes the anticipated movement through the various tests. The extent to which advancement through the hierarchy is required would depend on two factors:

- results from tests in the previous level (e.g., if there is acute toxicity, proceed to level II; if there is no acute toxicity, no further testing will be required unless environmental chemistry suggests persistence, in which case reproductive tests should be performed); and
- anticipated use patterns (e.g., if use does not indicate repeated human exposure, tests are not required for long-term toxicity, teratogenicity, or reproductive impairment).

The guidelines drafted under TSCA offer a basis for assessing potential risk associated with anticipated use of new chemicals. TSCA proposed recommendations do not include a rigid testing sequence. Manufacturers are given the responsibility for determining which tests should be performed to develop the essential "effects-data" with which scientifically sound assessments of risk may be made. Two sets of studies are described: reference studies and hazard evaluation studies. Reference studies are suggested for all new chemicals unless the manufacturer can show that some are not relevant to a particular substance (e.g., chemical intermediates and substances that are produced and used entirely in closed systems). The reference set includes:

- acute toxicity studies,
- subchronic toxicity studies,
- reproduction studies,
- teratogenic studies,
- mutagenic studies, and
- short-term predictive oncogenic studies.

Hazard evaluation studies include chronic toxicity and oncogenicity tests. Draft guidelines recommend these when data from reference studies are not adequate to predict potential toxicity of the new chemical, or when extensive hazard evaluation is considered essential.

The difference between FIFRA and TSCA proposed requirements can be summarized as follows:

- FIFRA establishes a set sequence of tests with trigger points for movement through the sequence;
- TSCA allows the manufacturer to determine the extent and sequence of testing deemed necessary for scientifically sound hazard evaluation; exemptions or departures from recommended tests may be permitted if adequate justification can be provided.

General descriptions of the tests specified in both guidelines are presented in Table 3.4.

Tests within the FIFRA scheme proceed from basic laboratory tests to applied field tests. Beyond tests of acute toxicity, wildlife data are required only when patterns of use suggest exposure in the habitat. When use patterns indicate no anticipated exposure or a minimal level of risk, data from chronic tests are not required.

As with the hazard evaluation tests for wildlife exposure risk, the data required for assessing risk to domestic animals and humans also would depend largely upon anticipated use patterns and results of acute toxicity tests.

In addition to the FIFRA tests for animal toxicity evaluation, TSCA requires assessments of effects on plants, both monocotyledons and dicotyledons. The recommended tests determine effects on plant growth and seed germination. The TSCA guidelines concentrate testing efforts on assessment of risk to human health and do not consider risk to domestic or commercial animals.

Committee's Comments: The individual tests proposed for hazard assessment appear to be valid. However, there remain two basic problems in the FIFRA evaluation scheme:

1. Physical and chemical properties apparently are not used to determine the appropriate testing route for each chemical except for deciding whether reproductive toxicity tests should be performed; and
2. Use patterns combined only with results of acute toxicity tests are inadequate means for selecting further appropriate tests of new chemicals. For example, in this evaluation format chronic data are required for human health assessment only if use patterns suggest repeated exposure over a significant portion of the human lifespan and acute toxicity results are positive; similarly, teratogenicity and reproductive studies are necessary only if significant exposure to humans is expected. The problems associated with these two points are discussed in the following pages using PCBs as the test case.

The relative flexibility of the TSCA guidelines reflects the fact that no single, rigid testing scheme is suitable for the evaluation of all chemicals. In the opinion of this Committee, the flexibility is good. However, the decision to allow exemptions from the test requirements or to require additional testing should not be left to an individual but should be made by a group of experts with backgrounds in chemistry, toxicology, ecology, and agriculture.

ASSESSMENT OF PCB HAZARD

An analysis of PCB data (Appendix D) compiled following the criteria of FIFRA and TSCA proposed guidelines, leads to the conclusion that PCBs are persistent, and are likely to accumulate. PCBs do not appear particularly toxic for short-term exposure, but results are subject to interpretation (Roberts et al. 1978). Based on anticipated patterns of use (Table 3.1), one might expect PCB distribution in the environment and a potential risk for humans through use of lacquers and varnishes containing PCBs.

TABLE 3.4 FIFRA and TSCA Data Requirements for Product Hazard Evaluation (Spp - species)

TERRESTRIAL AND AQUATIC WILDLIFE:	FIFRA Tests	FIFRA Criteria	TSCA Tests (Reference Set)	TSCA Criteria
Acute Toxicity	LD_{50} - one spp each; avian, mammal, fish LC_{50} - 2 spp each; avian, mammal, fish	Product used outdoors; test in systems most affected by use patterns (i.e., aquatic or terrestrial exposure)	EC_{50} - Daphnia and 2 spp of invertebrates from different classes or phyla LC_{50} - 1 spp fish; preferred spp are rainbow trout or sunfish	Product expected to enter aquatic system or be dispersed by water
	Life Cycle - fish embryo/larval or invertebrate	LC_{50}<1mg/l and water concentration >0.01% of LC_{50}; use patterns result in water concentrations <0.01% of LC_{50} with significant reproduction effects; physical chemical accumulation; $t_{.5}>4$ days; use pattern expect continuous water exposure	LC_{50} - quail dietary	Product not significantly volatilized from feed prior to consumption
	Life Cycle - fish, preferred spp minnow (fresh water or estuarine)	If embryo/larvae tests indicate adverse effects; environmental concentration is 0.1% "no effect" level of embryo/larvae tests, if effects observed in mammalian or avian reproduction with low concentration	Life Cycle - Daphnia; fish embryo/juvenile	Product enters aquatic systems and is soluble or dispersed
	Organism toxicity and residue studies	Previous test results suggest adverse effects on non-target organisms; considered on case-by-case basis		
Reproduction	One generation reproductive studies-quail and mallard	Physical, chemical and environmental data indicate product is persistent; compound or metabolites bioaccumulate; use patterns suggest repeated avian exposure	See Human Health Effects	
Field Studies	Tests for direct or indirect poisoning, behaviorial impairment, reproductive effects and evidence of ecological disturbances	If acute & reproductive data indicate adverse effects		

DOMESTIC ANIMALS AND HUMAN HEALTH:

	FIFRA Tests	FIFRA Criteria	TSCA Tests Reference Set	TSCA Criteria
Acute Toxicity	Oral and inhalation studies - rats	Required for all new compounds; critical levels: no mortality-dose level 5mg/kg, 5 animals/sex	Same tests as FIFRA	Recommended for all new chemicals that require TSCA notification
	Dermal studies - rabbits	Required for all new compounds; critical level 5mg/l for 4 hours		
	Dermal and eye sensitization studies - rabbits	Required for all new chemicals	Same tests as FIFRA	Recommended for all new chemicals that require TSCA notification
	Delayed neurotoxicity - hens; test parent compound and metabolites	Required for all new chemicals	Not recommended in Reference Tests	
Subchronic Toxicity	Oral dose - 2 mammalian spp (1 rodent, 1 non-rodent)	Required if use patterns indicate repeated human exposure is expected	Same as FIFRA	Recommended for all new chemicals that require TSCA notification
	Dermal studies - 21 day, rabbit	Required if use patterns indicate repeated skin contact is anticipated		
	90 day, rabbit	Required if expect purposeful skin application, use will result in human dermal exposure or expected to be metabolized in different manner dermally from oral or inhalation routes	Dermal - 90 day studies, same as FIFRA	Recommended for all new chemicals that require TSCA notification
	Inhalation studies - rat	Required if acute inhalation results are positive and use patterns indicate repeated inhalation exposure at toxic levels	Same as FIFRA	Recommended for all new chemicals that require TSCA notification
	Neurotoxicity	Required if acute test results were positive or chemicals are of structurally related to known neurotoxins	Not recommended in Reference Tests	
Reproductive Toxicity	Two generation studies - choice of rat, mouse, hamster or rabbit; same spp to be used in teratogenic studies	Required if use patterns indicated significant human exposure	Three generation studies - protocol and spp same as FIFRA	Recommended for all new chemicals that require TSCA notification

TABLE 3.4 continued

DOMESTIC ANIMALS AND HUMAN HEALTH:	FIFRA Tests	FIFRA Criteria	TSCA Tests	TSCA Criteria
			Reference Set	
Teratogenic Toxicity	Studies done using same spp as reproductive toxicity tests	Required if use patterns indicate widespread exposure to human females	Same as FIFRA	Recommended for all new chemicals that require TSCA notification
Mutagenic Toxicity	Short-term studies to include assessment of gene mutations, chromosomal abnormalities, and primary DNA damage	Required if use patterns indicate significant exposure through skin or inhalation is expected before or during reproductive portion of the human lifespan, or if chemically related to a known mutagen	Same as FIFRA	Recommended for all new chemicals that require TSCA notification
			Hazard Evaluation Set	
Chronic Toxicity	Oral, dermal or inhalation studies - to determine dose-related morbidity and mortality, determine target organ, specify nature of toxic effects and spp variation & sensitivity - 1 rodent and 1 non-rodent	Required if acute results are positive and use patterns indicate repeated human exposure over a significant portion of the human lifespan	Same Tests as FIFRA - 2 spp	Recommended if predicted human exposure is great and if results from Reference Set tests (above) suggests potential human health hazard, or if anticipated effects cannot be predicted by tests in the Reference Set
Oncogenic Toxicity	Tests with continuous exposure to chemical for animals lifespan - 2 mammalian spp	Required if parent, metabolic or degradation products structurally related to a known carcinogen, have mutagenic effects, or subchronic effects that may lead to neoplastic changes	Continuous testing over animals lifespan; route of exposure to be the same as major anticipated human exposure - 2 mammalian spp	Recommended if predicted human exposure is great and if results from Reference Set suggest potential human health hazard, if Reference tests cannot predict anticipated effects, or if chemical is structurally similar to a known carcinogen
Metabolic Studies	General metabolic tests to identify significant metabolic products; dose-related absorption, binding properties with target organ macromolecules - same spp as chronic feeding and oncogenicity studies	Required whenever chronic data are required as support for feeding on oncogenicity testing	No metabolic studies recommended	
	Data from the above tests may be required for domestic animals if exposure to the chemical is anticipated for cats, dogs, cattle, pigs, sheep, etc. This exposure can be through direct application or through consumption of feed		No recommended tests for domesticated or commercially used animals	

The FIFRA guidelines for hazard assessment predicate sequential testing on two factors: (1) evidence for acute toxicity, and (2) use patterns as an indication of anticipated exposure. The TSCA requirements call for more extensive testing (i.e., reproduction, subchronic, etc.) without consideration of use patterns or results of other tests.

FIFRA Testing Results

The conclusion reached regarding the potential hazard posed to the environment by PCBs based on FIFRA requirements is questionable. Acute toxicity is observed <u>only</u> for some aquatic invertebrates, but these species are not required test animals in the FIFRA guidelines. Based on anticipated use patterns, the expected human and environmental exposure to the chemical would be primarily from protective coatings. Some risk would be anticipated in occupational situations, but these exposures could be carefully controlled. If the anticipated use had been limited to heat-exchangers and closed electrical equipment, the testing data might not have suggested potential hazard to human health or the environment. In such a case PCBs might have "passed" the toxic substance evaluation. The aggregate data presented in Appendix D and the "real world" experience of more than 40 years of use of this chemical indicate widespread environmental exposure and possible serious health problems. The discrepancy is due to the fact that the current FIFRA protocols for hazard evaluation relate the degree of required testing to proposed uses. The PCB experience provides a useful test of the appropriateness of relating test requirements to projections of chemical uses.

The question of whether projected use patterns give an adequate indication of environmental exposure is complex. The use patterns anticipated prior to manufacture clearly do not give an adequate impression of the hazards associated with accidental release to the environment during manufacture and use, or of the extended problems of environmental release from disposal of equipment containing PCBs. The expanded uses of PCBs after 1930, illustrated in Table 3.5, may suggest increased exposure through disposal of products containing PCBs, but these uses still do not suggest the

TABLE 3.5 Uses of Aroclor by Type

Current Uses (since 1970)	1016	1221	1232	1242	1248	1254	1260	1262	1268
Capacitors	X	X		X*		X*			
Transformers				X		X	X*		
Die casting -- Feneclor-79%Cl									
Former Uses									
Heat transfer				X					
Hydraulic/lubricants									
• Hydraulic fluids			X	X	X	X	X		
• Vacuum pumps					X	X			
• Gas-transmission turbines		X		X					
Plasticizers									
• Rubbers		X	X	X	X	X			X
• Synthetic resins					X	X	X	X	X
• Carbonless paper				X					
Miscellaneous									
• Adhesives		X	X	X	X	X			
• Wax extenders				X		X			X
• Dedusting agents						X			
• Inks						X	X		
• Cutting oils						X			
• Pesticide extenders						X			
• Sealants and caulking compounds						X			

*Discontinued use of these types.

SOURCE: Modified from IARC (in press).

actual problems that developed. These problems are illustrated by the following examples.

Sterility in Mink in North Central United States and Canada (Aulerich and Ringer 1977) In early 1960, mink ranchers began noticing reproductive complications and excessive kit mortality among their stock. An acute problem was evident by 1967, resulting in an unprecedented 80 percent increase in newborn mortality. Subsequent investigation indicated a strong relationship between kit mortality and the percentage of coho salmon in the mother's diet, as well as the duration of feeding with a salmon diet. Factors such as rancidity, pesticide contamination and mercury poisoning were suspected but were shown not to contribute to the problem.

Experiments with the mink diet were conducted, and the results indicated that diets of 30 percent salmon produced the reproductive problems. Further investigation ruled out salmon per se as the cause, because other species of Great Lakes fish caused similar adverse effects. The following evidence strongly implicated PCBs as the causative factor:

- high PCB residue levels of Lake Michigan coho salmon;
- sensitivity of mink to PCBs;
- similarity of reproductive complications in mink fed coho salmon and those fed diets contaminated with PCBs;
- similarity of clinical signs and lesions observed in mink that died while on diets containing coho salmon and diets containing PCBs; and
- accumulation of PCB residues in mink tissue.

Rice Oil Contamination in Japan (Kuratsune et al. 1972, World Health Organization 1976) In June 1968, Japanese patients suffering from chloracne began appearing at the Kyshu University Hospital. By October of that year the disease was of epidemic proportions, involving more than 1,000 people in 21 prefectures in the western section of Japan. The exposure factor common to these patients was consumption of rice oil produced in February, 1968. Investigation of the manufacturing process identified a leak in a heat-exchanger, which resulted in oil contamination with Kanechlor 400, a 48 percent chlorinated biphenyl. In addition to symptoms associated with chloracne, clinical signs included a rise in serum transaminases and alkaline phosphatase, respiratory symptoms and chronic bronchitis-like problems, decrease in conduction velocity in peripheral sensory nerves, diminished growth in young males, and pigmented skin on newborns.

Between February and March of 1968, a disease occurred in chickens in the same section of Japan. More than 400,000 animals died. Clinical signs included labored breathing, droopiness, and decreased egg production. Autopsy indicated subcutaneous edema, hydropericardium, ascites, pulmonary edema, muscular ecchymosis, and a mottled appearance of the liver. Investigations revealed that the chicken feed had been prepared with the rice oil contaminated with Kanechlor 400. Subsequent study of the Kanechlor indicated large amounts of chlorodibenzofurans present. The extent of this chemical's involvement in the Yusho symptoms is not fully resolved.

<u>Sewage Sludge Problem in Bloomington, Indiana (McCloskey et al. 1978)</u> Residents of Bloomington, Indiana, had been using sewage sludge in fertilizing their gardens until high concentrations of PCBs were reported in the sludge in 1976. An investigation to locate the source revealed that a local electrical firm was using PCBs as a dielectric in capacitors. The company had been discharging wastes contaminated with PCBs into the municipal sewage system. By January 1976, sludge samples had mean PCB levels of 479 mg/kg and samples of soil treated with the sludge had mean concentrations of 17 mg/kg.

Health officials conducted clinical evaluations on workers in the electrical firm, their families, and Bloomington residents who had used the PCB-contaminated sludge for fertilizer. Serum PCB levels, alteration in hepatocellular function, and plasma triglyceride levels were examined. Although mean serum levels were relatively high compared to U.S. population groups (workers 72 mg/kg; families 34 mg/kg; residents 18 mg/kg; U.S. population 5 mg to 20 mg/kg) no correlation could be detected between the serum levels and the number of years using sludge, total pounds used, or time since the last use. Serum levels of selected liver enzymes increased significantly with PCB levels in serum as did plasma triglyceride levels.

Effluent from General Electric Plants in New York(Horn et al. [in press]) The Hudson River had been contaminated between 1942 to the mid 1970s by PCB effluent from General Electric manufacturing plants at Fort Edward and Hudson Falls, averaging about 14 kg per day. The long-term contamination has polluted virtually the entire river bed for many miles downstream. Bottom sediment below Fort Edward reportly contains 540 μg/kg to 2,980 μg/kg. The average concentration of PCBs in bass, perch, and sucker is 62 mg/kg to 135 mg/kg.

Contamination of Fish Meal at a Ralston Purina Plant (Jaroslovsky 1978) In 1977, a poultry firm began noticing high chemical levels in tissue samples taken from their chickens. The feed supplier was asked to help in pinpointing the cause, and the culprit was found to be fish meal contaminated by PCBs that had been used in formulating the chicken feed.

The fish meal had been prepared and stored in Puerto Rico at a Ralston Purina plant warehouse where two electrical transformers containing PCBs were stored also. Contamination of the meal occurred during a fire in April 1977, which damaged the electrical equipment and allowed PCBs to leak out. Water from fire hoses evidently mixed with the PCB fluid and soaked into the stored fish meal. The contamination was not detected and the fish meal was subsequently sold. As a consequence, 400,000 chickens and 15,000 dozen eggs were destroyed. Some of the contaminated feed did not actually contain the PCB fish meal, but the feed had been processed with the same machinery as the fish meal.

It is clear from these illustrations that projected patterns of use may not provide an adequate indication of the eventual distribution of PCBs in the environment or the extent of the hazard this distribution can pose. These incidents were the result of manufacturing and disposal procedures--factors not considered in proposing guidelines for testing new chemicals.

Toxic effects of PCBs in workers exposed in their occupations were noted as early as 1937 and might have been anticipated even earlier from the effects of chlorinated naphthalenes detected during World War I (Jacobs 1949). Had the awareness of toxic chemicals then been as widespread as it is currently, these incidents would have been more than sufficient to have curtailed the manufacture and use of PCBs or, at least, to have demanded detailed testing that would have revealed the chemical's chronic and more subtle toxic effects.

TSCA Testing Results

Physical and chemical properties of PCBs indicate the tendency of these compounds to persist as well as their high bioaccumulation potential. The data compiled following the Reference Set requirements and the proposed use of PCBs in protective coatings, which would result in some exposure of humans, justify the requirement for Hazard Evaluation Studies. These studies probably would detect at least some of the toxic effects of PCBs. However, it should be noted that most of the chronic tests proposed to assess human health hazards require use of species (rodent plus one other mammal) that would not adequately reflect the human toxic response to PCB exposure. The common laboratory test animals (rats and dogs) do not always respond to PCB exposure in the same manner as humans, and Table 3.6 suggests that no one species exhibits "typical" human symptoms of PCB poisoning except in the case of reproductive tests. Thus proper choice of the test species becomes important when evaluating the potential hazard of any substance.

Although PCBs have been detected in the U.S. human population (Kutz and Strassman 1975, Humphrey 1977), there have been no documented cases of PCB intoxication of humans other than those related to industrial exposure. Effects, when detectable, have been subtle rather than overt. For the most part, the proposed sets of tests for FIFRA and TSCA are designed to detect only gross intoxication. Subtle effects are not considered at all in the scheme proposed for TSCA and only peripherally in those drafted by FIFRA.

RECOMMENDATIONS

Perhaps the two most important properties of a chemical that relate to potential hazards to health and the environment are its persistence and lipophilicity. "High scores" in those tests that identify these properties should alert officials to the inevitability of environmental contamination and health problems. It

TABLE 3.6 Summary of Chronic Toxic Effects of PCBs

Test	Effects
Chronic Feeding	
Aquatic Species	Threshold effects in egg hatchibility of vertebrates and invertebrates at levels of 2-5 μg/l
	Embryo toxicity evident at 50 μg/l
Terrestrial Species	Mouse - some liver change with exposure to high chlorine containing products, 300-500 μg/g
	Rat - some liver changes, minimal reproductive effects, 100-500 μg/g
	Monkey - Yusho symptoms, altered reproduction cycles, hyperplastic gastritis and ulceration, 2.5-5 μg/g
	Chicken - some morphologic deformity, reproduction decline, subcutaneous edema, 20-50 μg/g
	Mink - dose response relationship in growth and reproduction, 10 μg/g
	Pelican - some hepatocellular changes, 100 μg
	Dogs - reduced growth, some liver changes, 100 μg
	Wildfowl - some reproduction changes, varies with species, 50-200 μg/g
Teratogenicity	Effects seen in avian species, 50-200 μg/g
Mutagenicity	Chromosomal abnormalities - negative results
	Dominant lethal mutations - negative results
	Ames test - 1221, 4 chlorobiphenyl significantly mutagenic
Oncogenicity	High chlorinated compounds produced tumors in rats and mice, relationship with PCB not always clear

should be emphasized, however, that evidence of nonpersistence or nonlipophilicity does not suggest lack of hazard potential.

Physical, chemical, and environmental chemistry data are sufficient to determine persistence and lipophilic properties of chemicals and can be used to predicate potential bioaccumulation and environmental distribution. Therefore, the Committee recommends a multitiered evaluation scheme as illustrated in Figure 3.1. This scheme requires first:

- chemical identification of the parent compound, impurities, metabolites, and degradation products;
- determination of physical and chemical properties; and
- assessment of environmental and bioscreening (short-term toxicity) test results for parent compound and metabolites.

At this point, if the data indicate that the chemical is highly persistent or lipophilic, or both, its development for large-scale use should be scrutinized carefully and possible alternatives considered. On the basis of the scrutiny, the chemical may be either rejected or subjected to additional testing. If the data indicate that the chemical is not highly persistent or lipophilic but there are indications of potential hazards, the chemical should be subjected to additional testing. If there are no indications of potential hazards, the chemical may be approved for manufacture and use.

Additional assessment of hazard potential for all suspect substances should include:

- development of a mass balance model of the potential movement of the substance through the environment using data regarding physical and chemical properties and anticipated source function parameters;
- identification of "problem areas" or sites of potential chemical accumulation (e.g., agricultural lands, urban areas, marine or freshwater sources, or combinations of these);
- specification of tests to assess toxic effects within the "problem areas."

Test guidelines should be formulated specifying representative species for each environmental medium and

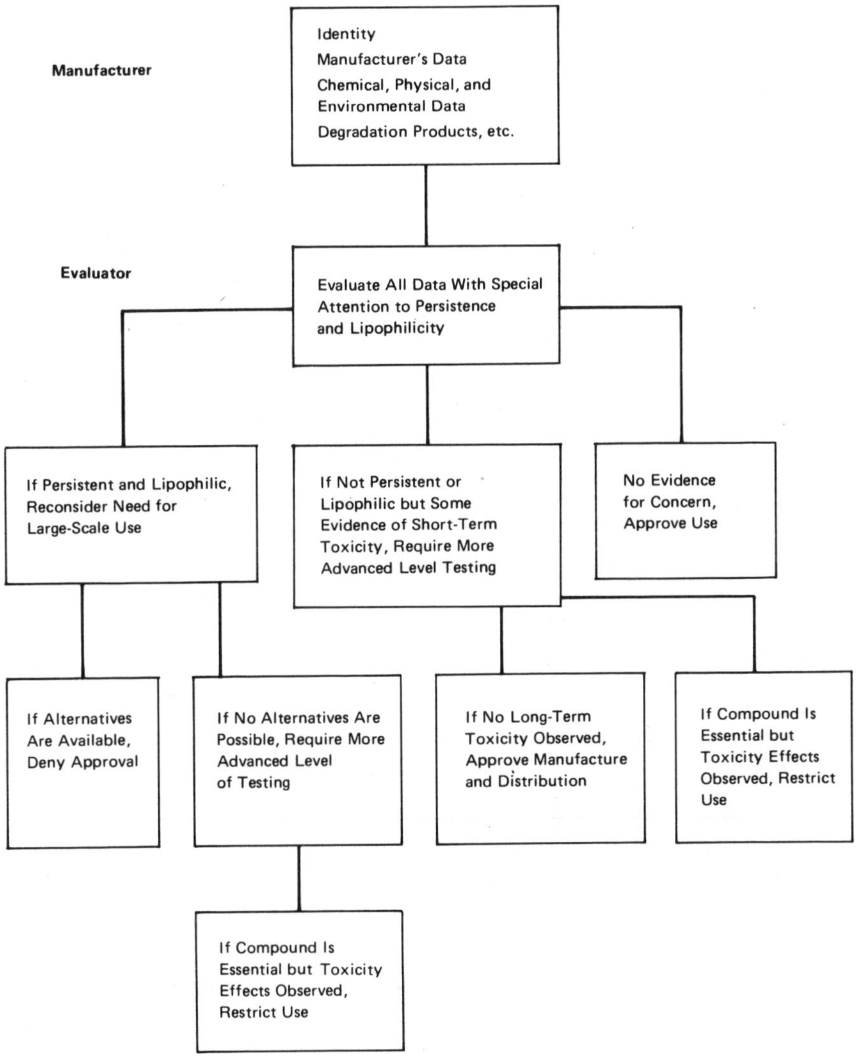

FIGURE 3.1 Evaluation scheme for assessing potential hazards to health and the environment.

those species most appropriate for assessing toxicity to humans. Specified tests should include assessment of:

1. reproductive alterations;
2. pathological effects; and
3. behavioral and learning impairment.

Using such an approach, industry could concentrate its testing efforts on identifying effects in specific ecosystems. For example, if the model predicts primary chemical accumulation in the sediment of rivers and lakes, the manufacturer could concentrate on assessment of hazards within these systems, using quick screening tests for other media when results indicated possible contamination of land or air. Monitoring after a product has been marketed should be required to keep industry and government alert to unexpected accumulations, and additional tests and the implementation of appropriate control measures could be demanded as necessary.

REFERENCES

Aulerich, R.J. and R.K. Ringer (1977) Current status of PCB toxicity to mink, and effect on their reproduction. Arch. Environ. Contam. Toxicol. 6:279-292.

Chiu, C.T., V.H. Freed, D.W. Schmedding, and R.L. Kohnert (1977) Partition coefficient and bioaccumulation of selected organic chemicals. Environ. Sci. Technol. 11:475-478.

Conservation Foundation (1978) Approaches for Developing Testing Guidelines Under the Toxic Substances Control Act. (Unpublished)

Horn, E.G., L.J. Hetting, and J. Tofflemire (In press) The problem of PCBs in the Hudson River system. The scientific basis for public control of environmental health hazards, symposium, 1976. New York Acad. Sci. Science Week. (To be published Spring 1979)

Humphrey, H.E.B. (1977) Evaluation of Changes in the Level of Polychlorinated Biphenyl in Human Tissue. Final Report on FDA Contract No. 223-73-2209. Michigan Department of Public Health.

International Agency for Research on Cancer (IARC) (In press) Monographs on the evaluation of the carcinogenic risk of chemicals to humans 18:43-103. (Due for release in or after 1979)

Jacobs, M.B. (1949) The Analytical Chemistry of Industrial Poisons, Hazards, and Solvents. New York: Interscience Publishers, Inc.

Jaroslovsky, R. (1978) Contamination of fish meal with PCB sparks FDA study of mishap at a Ralston Purina plant. Wall St. J. 46, July 12.

Kuratsune, M., T. Yoshimura, J. Matsuzaha, and A. Yamaguchi (1972) Epidemiological study of Yusho, a poisoning caused by ingestion of rice oil contaminated with a commercial brand of polychlorinated biphenyls. Environ. Health Perspect. 1:119-128.

Kutz, F.W. and S.C. Strassman (1975) Residues of polychlorinated biphenyls in the general population of the United States. Pages 138-143, Proceedings of National Conference on Polychlorinated Biphenyls, November 19-21, 1975, Chicago, Illinois. Report No. EPA-560/6-75-004. Washington, D.C.: U.S. Environmental Protection Agency.

McCloskey, F.X., T.O. Middleton, R. Schneider, J. VanFleet, W.T. Paynter, R.D. Telle, and C.J. Glueck (1978) Polychlorinated biphenyl exposure - Indiana. Morbidity and Mortality Weekly Report 27(12):99-100.

Moriguchi, I., Y. Kanada, and K. Komatsu (1976) Van Der Waals volume and the related parameters for hydrophobicity in structure-activity studies. Chem. Pharmacol. Bull. 24:1799-1806.

Nisbet, I.C.T. (1976) Criteria Document for PCBs. Report No. PB-225 397. Washington, D.C.: U.S. Department of Commerce.

Oishi, S. and K. Hiraga (1978) Is a mixture of polychlorinated dibenzofurans an inducer of hepatic porphyria? Food Cosmet. Toxicol. 16(1):47-48.

Oishi, S., M. Morita, and H. Fukuda (1978) Comparative toxicity of polychlorinated biphenyls and dibenzofurans in rats. Toxicol. Appl. Pharmacol. 43:13-22.

Penning, C.H. (1930) Physical characteristics and commercial possibilities of chlorinated diphenyl. Ind. Eng. Chem. 22(11):1180-1182.

Roberts, J.R., D.W. Rodgers, J.R. Bailey, and M.A. Rorke (1978) Polychlorinated Biphenyls: Biological Criteria for an Assessment of Their Effects on Environmental Quality. Associate Committee on Scientific Criteria for Environmental Quality. Ottawa, Canada: National Research Council of Canada.

Smith, J.H., W.R. Mabey, N. Bohonos, B.R. Holt, S.S. Lee, T.W. Chou, D.C. Bomberger, and T. Mill (1977) Environmental Pathways of Selected Chemicals in Freshwater Systems. Part I: Background and Experimental Procedures. Report No. EPA-600/7-77-113. Washington, D.C.: U.S. Environmental Protection Agency.

U.S. Environmental Protection Agency (1978a) Draft Guidelines for Testing Toxic Substances. Office of Toxic Substances. (Unpublished)

U.S. Environmental Protection Agency (1978b) FIFRA Proposed Guidelines for Registering Pesticides in the United States. Subpart B-Introduction, Subpart D-Chemistry Requirements, Subpart F-Hazard Evaluation: Wildlife and Aquatic Organisms, Subpart F-Hazard Evaluation: Humans and Domestic Animals. Federal Register 43(132):29696-29741 and 43(163):37351-37403.

World Health Organization (1976) Polychlorinated Biphenyls and Terphenyls. Environmental Health Criteria 2. Geneva: World Health Organization.

Zitko, V., D.J. Wildish, O. Hutzinger, and P.M.K. Choi (1973) Acute and chronic oral toxicity of chlorinated dibenzofurans to salmonid fishes. Environ. Health Perspect. 2:187-189.

APPENDIX A

AREAS, VOLUMES, AND MISCELLANEOUS QUANTITIES FOR CHAPTER 1 CALCULATIONS[1]

[1]The values in this table were used in calculating environmental burdens of PCBs.

TABLE A.1 Areas, Volumes, and Miscellaneous Quantities

U.S. Area	Area of Drainage Basin (km^2)	Agricultural Land (km^2)[b]	Discharge of Water (km^3/yr)	Discharge of Suspended Sediment (10^6 metric tons/yr)[c]	Area of Sediments Origin (km^2)[c]	Sediment Yield (metric tons/km^2)[c]
Atlantic[a]	1,361,664	1,100,288	321.258	12.886	743,760	17.3
Gulf of Mexico	4,315,392	3,652,766	788.99	343.080	4,504,528	76.1
Pacific[a]	2,108,672	1,523,256	445.97	84.872	1,637,942	54.9
Total	7,785,728	6,276,310	1,556.318	445.838	6,886,230	

	Streams[d]		Reservoirs having Capacity > 5000 acre-feet		Lakes > 10 mi^2[d]	
U.S. Area	Surface Area (km^2)	Volume of Water (km^3)	Surface Area (km^2)	Volume of Water (km^3)	Surface Area (km^2)	Volume of Water (km^3)
Atlantic	21,299	106.5	11,655	48.393	163,264	11,623
Gulf of Mexico	25,728	128.6	35,274	227.5	12,966	83.8
Pacific	5,458	27.2	8,376	153.761	3,080	341.8
Total	52,485	262.3	55,305	429.854	177,310	12,048.6

	Sediments		
U.S. Area	Volume of Stream Sediments Containing PCBs (km^3)[e]	Volume of Reservoir Sediments Containing PCBs (km^3)[f]	Lake Sediments Containing PCBs (km^3)[g]
Atlantic	4.472	5.827	16.326
Gulf of Mexico	5.402	17.637	1.297
Pacific	1.146	4.188	0.308
Total	11.020	27.652	17.931

[a]In general, our boundaries coincide with those in the Water Resources Council's report. Two differences are (1) with Atlantic are included the Souris-Rainy-Red drainage into the Hudson Bay and the Great Lakes area, which drains via the Gulf of St. Lawrence. The value for area in the Great Lakes Region excludes the water-surface area of the U.S. portion of the Great Lakes. (2) The Pacific column includes the area of the Great Basin. The data in the Water Resources report did not enable us to

separate the streams that are unified in these drainages from the usage in the Water Resources Council's book.

[b]Agricultural land includes cropland, both that used for crops and that left idle and fallow; permanent pasture; and forest and woodland. In the Water Resources report, these areas were tabulated in acres. They were converted to square kilometers by finding the proportion of the total land area, which is given both as 10^3 acres and square miles, and then by converting this to square kilometers (using 2.56 km^2 = 1 mi^2). The figures shown are the areas as projected for 1980.

[c]The figures for discharge of suspended sediment from the three areas are from Curtis, Culbertson, and Chase, 1973, U.S. Geological Survey Circ. 670, Table 3, p. 11. The reasons for the discrepancies between the areas of the drainage basins between U.S. Geological Survey Circ. 670 and the figures shown here from the Water Resources Council are not known.

[d]Part of the information required for this study included the surface areas of lakes and rivers within the United States, the volumes of water in the lakes and rivers, volumes of sediments containing PCBs, and the concentration of PCBs in the water. For the Great Lakes area, only the U.S. portion was included.

In contrast to the large array of data on the areas of drainage basins, lengths of river systems, and quantities of water and suspended sediment discharged, data on the areas and volumes for the present purpose are extremely scarce.

The basis for the areas of the water bodies used in this study is the 1940 census compilation of "Inland Water of the United States." These figures are summarized state by state in the National Atlas, but in the Atlas, the definition of inland water was not included. A tabulation with the definition, but attributed incorrectly to the 1960 census, appears on page 181 of the Water Encyclopedia. In the 1940 census, inland water was defined as lakes, reservoirs and ponds having an area of 40 acres or more, and streams, sloughs, and channels having widths of 1/8 mile and greater.

From the total area of inland water were subtracted the areas of the following: (1) lakes having areas of 10 square miles or more (from D.K. Thomas, ed., 1970, The Water Encyclopedia, p. 122-124); (2) large saline lakes (from Water Encyclopedia, p. 128-129); and (3) reservoirs having a usable capacity of 5000 acre feet or more (from Water Encyclopedia, p. 410).

The category for streams does not include those narrower than 1/8 mile but does include lakes and ponds having areas of greater than 40 acres but less than 10 square miles.

The areas in the "stream" column are accurate to the extent that the cumulative areas of lakes and ponds smaller than 10 square miles but larger than 40 acres equals the cumulative area of streams narrower than 1/8 of a mile.

In a few states, the subtraction of the areas of the reservoirs after reducing "inland water" by subtracting the areas of fresh water lakes greater than 10 square miles and the areas of saline lakes yielded a negative number. We took this to mean that in those states, large reservoirs had been built since 1940. We have not tried to check this independently. Where a negative number would have resulted, we did not subtract the areas of the reservoirs.

To get the mean volume of water in the rivers, we assumed that the mean depth of streams is 2.5 meters.

[e]The volume estimate for PCB-contaminated sediments has been computed by assuming that 70 percent of the "streams" area is underlain by sediment and that of this, the top 0.3 meters contains PCBs. This figure is based on the supposition that the data from the Upper Hudson River can be applied to all other streams.

[f]The volume of PCB-containing reservoir sediments was computed using an average thickness of 0.5 meters times the areas of the reservoir.

[g]The volume of PCB-contaminated lake sediments was computed using an average thickness of 0.1 meters times the areas of the lakes.

REFERENCES

Curtis, W.F., J.K. Culbertson, and E.B. Chase (1973) Fluvial-sediment discharge to the oceans from the conterminous United States. U.S. Geological Survey Circ. 670. Washington, D.C.: U.S. Geological Survey.

Todd, D.K., ed. (1970) The Water Encyclopedia. Port Washington, N.Y.: Water Information Center, Inc.

U.S. Department of the Interior Geological Survey (1970) The National Atlas of the United States of America. Washington, D.C.: U.S. Government Printing Office.

U.S. Water Resources Council (1968) The Nation's Water Resources: The First National Appraisal. Washington, D.C.: U.S. Government Printing Office.

APPENDIX B

COMPUTATION OF INCREMENTAL COSTS PER KILOGRAM CONTROLLED FOR EXISTING AND PROPOSED EPA REGULATIONS

This Appendix provides computations for each of the average cost figures of Table 2.1. The sources for the data are Versar, Inc. (1977), Westin et al. (1978), and Steven Malkenson, Office of Planning and Management, U.S. Environmental Protection Agency.

A. Drain and Incinerate Askarel Fluids from PCB Transformers.

Each transformer contains approximately 1,364 kg of fluid of which 977 kg are PCBs. Draining is 90 percent effective, yielding 1,228 kg of fluid (880 kg of PCBs) for handling and processing. A cost of $536 or $0.61/kg of PCB is derived based on the following:

Transportation charge, $0.07/kg	$ 81
Incineration charge, $0.33/kg	405
Handling charge per unit	50
	$536

B. Shred and Incinerate High Voltage Capacitors.

If each capacitor weighs 55 kg and contains 11 kg of PCBs, the cost-per-kg destroyed is $3.86.

Handling and transportation charge, $0.13/kg	$ 7.15
Incineration charge, $0.55/kg	30.35
Handling charge per unit	5.00
	$42.50

C. Shred and Incinerate Low Voltage Capacitors.

If each capacitor weighs 9.1 kg and contains 1.6 kg of PCBs, a cost of $7.19 or $4.49/kg is derived.

Handling and transportation charge, $0.13/kg	$1.18
Incineration charge, $0.55/kg	5.00
Handling charge per unit	1.00
	$7.19

D. Flush and Drain Transformers and Incinerate Fluid.

Each transformer uses 1,022 liters of solvent at $0.10/l. Contaminated solvent weighing 1,350 kg must be shipped and incinerated at a cost of $0.40/kg (see A). If the process removes 90 percent (or 88 kg) of the PCBs remaining after draining, the cost-per-kg of PCBs destroyed will be $8.50/kg.

Solvent	$108.00
Transportation and incineration	540.00
Handling charge per unit	100.00
	$748.00

E. Place Transformer Carcasses in Chemical Waste Landfill.

Disposal costs are estimated at $3/ft³ or $0.07/kg with transportation costs of $0.04/kg. The average weight of the transformer is 2,955 kg and cost per unit is $325. If each unit contains 9.8 kg of PCBs, the cost-per-kg of PCBs immobilized will be $33.16.

F. Replace Electromagnets Containing PCBs and Incinerate Fluid.

Westin et al. (1978) estimates that 200 units containing 500 kg to 750 kg of PCBs per unit would require fluid replacement at a total cost of $2.8 million to $3.4 millon or $18.67/kg to $35.00/kg.

G. Replace Motors Containing PCBs in Mining Equipment.

The following estimates are based on Westin et al. (1978). The regulations would force the rebuilding of 652 loader motors containing 20 kg of PCBs each at a cost of $3,100 or $155/kg. Equipment in service contains a total of 1,145 kg of PCBs. Costs of premature scrapping or replacement of these machines would range from $640,000 to $2,240,000 or $560/kg to $1,955/kg.

H. Drain and Incinerate Mineral Oil Transformer Fluids if Contaminated by More Than 50 mg/kg PCBs.

Westin et al. (1978) estimates that the cost of testing transformer fluids and subsequent incineration if necessary would range from $613 million to $769 million. It is estimated that transformer fluids may be contaminated by 175,000 kg of PCBs. If all of the PCBs were detected and removed, the average cost would be $3,500/kg to $4,400/kg. However, this calculation may underestimate the average cost of destruction of PCBs. If in the absence of regulation, contaminated fluids were normally used as utility boiler fuel, some fraction of the PCBs would be destroyed in any event. If 95 percent of the PCBs are destroyed when combustion occurs in utility boilers, then the average destruction attributable to the regulation is only 8,750 kg (0.05 x 175,000), in which case the average cost of destruction of PCBs is $70,000/kg to $88,000/kg. It should be noted however that the evidence on effective destruction of PCBs in utility boilers is scanty. If for example the destruction rate were as low as 50 percent, the average cost in Table 2.1 would be lower by an order of magnitude.

I. Flush and Replace Contaminated Hydraulic Fluids in Die-Casting.

Westin et al. (1978) estimate a cost per machine of $14,600 and an average PCB contamination of 0.11 kg/machine or $133,000/kg. However, there may be substantial variation in the degree of contamination in different machines. True average costs would be much lower for a policy directed only at the most contaminated machines, but would rise as the policy was extended to machines with lesser contamination.

REFERENCES

Versar, Inc. (1977) Microeconomic Impacts of the Proposed Marking and Disposal Regulations for PCBs. NTIS Report No. PB-267 833/2WP. Springfield, Va.: National Technical Information Service.

Westin, R.A., L. Fourt, D. Berkey, and B. Woodcock (1978) Microeconomic Impacts of the Proposed "PCB Ban Regulations." National Technical Information Service, NTIS Report No. PB 281-881/3WP. Springfield, Va.: Versar, Inc.

APPENDIX C

THE BASIS FOR THE VALUE OF STATISTICAL LIFE

A number of approaches to assigning monetary values to the loss of life have been proposed or used in the literature on the economics of health and safety. The approaches can be broadly categorized by the way they determine values, i.e., either according to individual preferences (willingness to pay) or according to resource or opportunity costs. The following is a brief evaluation of the two approaches. (For more extensive discussions of alternative concepts of the value of life or safety, see Schelling 1968, Mishan 1971, Acton 1973, and Jones-Lee 1976.)

Willingness to Pay

Conceptually, the more attractive approach is to value increases in longevity or reductions in the probability of death due to accident or illness according to what an individual is willing to pay to achieve them. Individuals often act as if life expectancy were like economic good, that is, they are willing to trade life expectancy for goods and services that they value more highly, or vice versa. Individuals make decisions that can result in reductions in life expectancy or increased probability of death in return for increases in income or other goods and services, and perceive themselves to be better off because of these choices. Examples would be accepting risky or hazardous jobs because of higher wages, or travelling by air instead of on the ground because air travel is quicker and more convenient.

As these examples make clear, the question is not how much an individual would be willing to pay to avoid certain death tomorrow, but rather how much he would be willing to pay to achieve a small decrease in the probability of death during a given period. This willingness to pay can be translated into a more useful figure for policy evaluation, namely, the value of statistical life or the value of a statistical death avoided. Suppose that in a group of 1,000 similar individuals, each would be willing to pay $1,000 for a policy that would reduce the probability of death by 0.01. This policy is a form of collective good for the individuals involved. The benefit to the group is found by adding across all individuals. Thus, the aggregate willingness to pay would be $1 million (1,000 x $1,000), and the expected number of deaths avoided would be 10. The statistical value per life would be $1 million divided by 10 or $100,000.

Two approaches to the empirical measurement of willingnesss to pay have been used in the literature. One is to observe market transactions where individuals actually purchase or sell changes in their risk levels. For example, if wage differentials among occupations are related to differences in occupational risk levels, these differences might be interpreted as reflecting, at the margin, the individuals' trade-off between risk/safety and money. The other approach is to conduct surveys that ask individuals a series of questions about hypothetical situations involving risk/money trade-offs. If the questions are carefully designed, and if individuals are truly capable of predicting how they would act if placed in these hypothetical situations, their answers may reveal the monetary values they attach to reductions in risk.

There have been two major studies using wage rates to estimate willingness to pay. Both studies are based on the same wage data but they employ different sets of occupational risk data and thus present quite different results. Thaler and Rosen (1976) concluded that the statistical value of life lies between $140,000 and

$260,000, the best estimate being $200,000. The figures are in sharp contrast to those of Smith (1976) whose estimates ranged between $1.5 million and $2.6 million.

There are also differences in the results of two efforts to use the survey approach to obtaining values for statistical life. Acton (1973) obtained 36 responses from a stratified random sample of residents in the Boston area. The survey instrument contained a number of questions about attitudes and value judgments with respect to efforts to save lives in emergency situations. The question of greatest interest concerned the willingness of the individual to pay for a program of emergency coronary care that would reduce the probability of his death from heart attack. Two different forms of the question implied values for statistical life of $28,000 and $43,000. Jones-Lee (1976) asked a similar small sample of individuals several questions about their willingness to accept higher air fares to travel on lines that had lower probabilities for a fatal crash. The value of statistical life implied by the respondents was about $5 million. The disparities suggest that substantially more research is required before reliable estimates of the value of life can be obtained with such techniques. But the results may suggest order of magnitude bounds on appropriate figures.

Opportunity Cost/Human Capital

The more common approach to the valuation of life is the opportunity cost, productivity, or human capital approach which values each life lost at the present level of the earnings that an individual could have expected had the death not occurred. The approach is based on the assumption that earnings reflect the individual's marginal productivity, or the individual's contribution to total economic output.

There are three criticisms of this approach. First, it has no necessary relationship to individual willingness to pay. Second, it allows no role for the probabilistic nature of death and the avoidance of death in the health and safety areas, or for different individual attitudes or preferences toward risk and its avoidance. By definition an individual could pay no more than the current value of his earnings to avoid certain death, although the statistical value of life based upon willingness to pay for small probability changes could be several times the discounted expected earnings stream. Third, the implicit judgment underlying the approach is that an individual is worth what he or she does, or that productivity is the measure of worth.

Because of variation in patterns of earnings over a lifetime and differences among individuals in their experiences in the labor market (including the factor of discrimination), values derived from this approach relate crucially to age, sex, and race. For example, using a 6 percent discount rate and 1972 patterns of earnings, saving the life of a white male between 30 and 34 years of age would prevent the loss of about $180,000 in discounted lifetime earnings. For a nonwhite woman of the same age, the earnings loss prevented would be only about $70,000 (Cooper and Rice 1976). Some benefit-cost analyses using productivity values have attempted to adjust for these factors. Others have used averages for the population as a whole, while still other studies have added estimates of the direct costs of illness as represented by such items as doctors' fees and hospital bills.

Judgmentally Derived Shadow Prices

As indicated above, empirical estimates of the value of statistical life vary within two orders of magnitude, and any of them can be questioned on statistical and conceptual grounds. An alternative to the empirical estimate is the judgmentally determined value or range of values. There are several instances of this approach in the literature on the evaluation of health and safety benefits.

The earliest official use of an explicit value of life in the analysis of proposed governmental strategies appeared in the RECAT report (Office of Science and Technology 1972) in which a life was valued at $140,000 in estimating the costs of fatalities associated with automobile accidents. The figure represents the average foregone participation in the personal income of the nation; it is the product of per capita personal income in 1970 times the estimated reduction of the expected life span of the average accident victim. Another use of a value of life occurred in a report by the National Research Council (1974) where the value of improved health to be expected from reducing gaseous residual discharges from automobiles was arrived at on the basis of $200,000 per death avoided. Another report by the National Research Council (1978) used a range of $100,000 to $1,000,000 per death avoided. Most of the more broadly based empirically derived estimates cited above reside within this range. For these reasons we chose to use this range for the calculations in the text.

REFERENCES

Acton, J.P. (1973) Evaluating Public Programs to Save Lives: The Case of Heart Attacks. Santa Monica, Calif.: Rand Corporation.

Cooper, B.S., and D.P. Rice (1976) The economic cost of illness revisited. Soc. Security Bull. 39(2):21-36.

Jones-Lee, M.W. (1976) The Value of Life: An Economic Analysis. Chicago: University of Chicago Press.

Mishan, E.J. (1971) Evaluation of life and limb: A theoretical approach. J. Political Econ. 79(4):687-705.

National Research Council (1974) Air Quality and Automobile Emission Control. Volume 4: The Costs and Benefits of Automobile Emission Control. A report by the Coordinating Committee on Air Quality Studies, Commission on Natural Resources, National Academy of Sciences/National Academy of Engineering for the Committee on Public Works, U.S. Congress Senate. Serial No. 93-24. 93rd Congress, 2nd Session.

National Research Council (1978) Chloroform, Carbon Tetrachloride, and Other Halomethanes: An Environmental Assessment. Committee for Scientific and Technical Assessments of Environmental Pollutants, Environmental Studies Board, Commission on Natural Resources. Washington, D.C.: National Academy of Sciences.

Office of Science and Technology (1972) Cumulative Regulatory Effects on the Cost of Automotive Transportation. Report of the Ad Hoc Committee. Washington, D.C.: U.S. Government Printing Office.

Schelling, T.C. (1968) The life you save may be your own. _In_ Problems in Public Expenditure Analysis. Edited by S.B. Chase, Jr. Washington, D.C.: The Brookings Institution.

Smith, R.S. (1976) The Occupational Safety and Health Act: Its Goals and Its Achievements. Washington, D.C.: American Enterprise Institute for Public Policy Research.

Thaler, R.H. and S. Rosen (1976) The value of saving a life: Evidence from the labor market. In Household Production and Consumption. Edited by N.E. Terleckyj. New York: Columbia University Press.

APPENDIX D

CHEMICAL AND TOXICITY DATA AS REQUIRED BY FIFRA AND TSCA GUIDELINES

CHEMICAL IDENTITY

Product Identity

The PCB product contains a mixture of chlorinated biphenyls; each biphenyl can be present as any one of a number of isomers. Table D.1 lists the chemical name and registry number for the most common (IARC, in press).

Synonyms for PCBs include chlorinated biphenyl, chlorinated diphenyl, chlorobiphenyl, polychlorinated biphenyl, and polychlorobiphenyl.

The following trade names have been used:

USA:	Aroclor	France:	Phenoclor	Italy:	Fenclor
	Chloretol				
	Dykanol	Japan:	Kanechlor	USSR:	Soval
	Inerteen		Santotherm		
	Noflamol				
	Pyranol	FRG:	Clophen		

The molecular formula and weight, percent chlorine and number of isomers are shown in Table D.2 (Durfee et

TABLE D.1 Chemical Abstracts Register Number for PCBs

Biphenyl

Chem. Abstr. Services Reg. No.: 92-52-4
Chem. Abstr. Name: 1,1'-biphenyl

2-Chlorobiphenyl

Chem. Abstr. Services Reg. No.: 2051-60-7
Chem. Abstr. Name: 2-Chloro-1,1'-biphenyl

4-Chlorobiphenyl

Chem. Abstr. Services Reg. No.: 2051-62-9
Chem. Abstr. Name: 4-Chloro-1,1'-biphenyl

2,2'-Dichlorobiphenyl

Chem. Abstr. Services Reg. No.: 13029-08-8
Chem. Abstr. Name: 2,2'-Dichloro-1,1'-biphenyl

2,3'-Dichlorobiphenyl

Chem. Abstr. Services Reg. No.: 25569-80-6
Chem. Abstr. Name: 2,3'-Dichloro-1,1'-biphenyl

2,4'-Dichlorobiphenyl

Chem. Abstr. Services Reg. No.: 34883-43-7
Chem. Abstr. Name: 2,4'-Dichloro-1,1'-biphenyl

4,4'-Dichlorobiphenyl

Chem. Abstr. Services Reg. No.: 2050-68-2
Chem. Abstr. Name: 4,4'-Dichloro-1,1'-biphenyl

2,2',3-Trichlorobiphenyl

Chem. Abstr. Services Reg. No.: 38444-78-9
Chem. Abstr. Name: 2,2',3-Trichloro-1,1'-biphenyl

2,3',4-Trichlorobiphenyl

Chem. Abstr. Services Reg. No.: none available
Chem. Abstr. Name: 2,3',4-Trichloro-1,1'-biphenyl

2,4,4'-Trichlorobiphenyl

Chem. Abstr. Services Reg. No.: 7012-37-5
Chem. Abstr. Name: 2,4,4'-Trichloro-1,1'-biphenyl

2',3,4-Trichlorobiphenyl

Chem. Abstr. Services Reg. No.: 38444-86-9
Chem. Abstr. Name: 2',3,4-Trichloro-1,1'-biphenyl

2,3,4,4'-Tetrachlorobiphenyl

Chem. Abstr. Services Reg. No.: 33025-41-1
Chem. Abstr. Name: 2,3,4,4'-Tetrachloro-1,1'-biphenyl

2,2',3,5'-Tetrachlorobiphenyl

Chem. Abstr. Services Reg. No.: 41464-39-5
Chem. Abstr. Name: 2,2',3,5'-Tetrachloro-1,1'-biphenyl

2,2',4,5'-Tetrachlorobiphenyl

Chem. Abstr. Services Reg. No.: 41464-40-8
Chem. Abstr. Name: 2,2',4,5'-Tetrachloro-1,1'-biphenyl

2,3',4,4'-Tetrachlorobiphenyl

Chem. Abstr. Services Reg. No.: 32598-10-0
Chem. Abstr. Name: 2,3',4,4'-Tetrachloro-1,1'-biphenyl

2,2',5,5'-Tetrachlorobiphenyl

Chem. Abstr. Services Reg. No.: 35693-99-3
Chem. Abstr. Name: 2,2',5,5'-Tetrachloro-1,1'-biphenyl

2,3',4',5-Tetrachlorobiphenyl

Chem. Abstr. Services Reg. No.: 32598-11-1
Chem. Abstr. Name: 2,3',4',5-Tetrachloro-1,1'-biphenyl

3,3',4,4'-Tetrachlorobiphenyl

Chem. Abstr. Services Reg. No.: 32598-13-3
Chem. Abstr. Name: 3,3',4,4'-Tetrachloro-1,1'-biphenyl

2,3,3',4,4'-Pentachlorobiphenyl

Chem. Abstr. Services Reg. No.: 32598-14-4
Chem. Abstr. Name: 2,3,3',4,4'-Pentachloro-1,1'-biphenyl

2,2',3,4,5'-Pentachlorobiphenyl

Chem. Abstr. Services Reg. No.: 38380-02-8
Chem. Abstr. Name: 2,2',3,4,5'-Pentachloro-1,1'-biphenyl

2,2',5-Trichlorobiphenyl

Chem. Abstr. Services Reg. No.: 37680-65-2
Chem. Abstr. Name: 2,2',5-Trichloro-1,1'-biphenyl

2,4',5-Trichlorobiphenyl

Chem. Abstr. Services Reg. No.: 16606-02-3
Chem. Abstr. Name: 2,4',5-Trichloro-1,1'-biphenyl

2,2',3,5',6-Pentachlorobiphenyl

Chem. Abstr. Services Reg. No.: 38379-99-6
Chem. Abstr. Name: 2,2',3,5',6-Pentachloro-1,1'-biphenyl

2,3,3',4',6-Pentachlorobiphenyl

Chem. Abstr. Services Reg. No.: 38380-03-9
Chem. Abstr. Name: 2,3,3',4',6-Pentachloro-1,1'-biphenyl

2,2',3',4,5-Pentachlorobiphenyl

Chem. Abstr. Services Reg. No.: 41464-51-1
Chem. Abstr. Name: 2,2',3',4,5-Pentachloro-1,1'-biphenyl

2,2',4,4',5-Pentachlorobiphenyl

Chem. Abstr. Services Reg. No.: 38380-01-7
Chem. Abstr. Name: 2,2',4,4',5-Pentachloro-1,1'-biphenyl.

2,2',4,5,5'-Pentachlorobiphenyl

Chem. Abstr. Services Reg. No.: 37680-73-2
Chem. Abstr. Name: 2,2',4,5,5'-Pentachloro-1,1'-biphenyl

2,3',4,4',5-Pentachlorobiphenyl

Chem. Abstr. Services Reg. No.: 31508-00-6
Chem. Abstr. Name: 2,3',4,4',5-Pentachloro-1,1'-biphenyl

2,2',3,4,4',5-Hexachlorobiphenyl

Chem. Abstr. Services Reg. No.: 35694-06-5
Chem. Abstr. Name: 2,2',3,4,4',5-Hexachloro-1,1'-biphenyl

2,2',3,3',4,6-Hexachlorobiphenyl

Chem. Abstr. Services Reg. No.: 38380-05-1
Chem. Abstr. Name: 2,2',3,3',4,6-Hexachloro-1,1'-biphenyl

2,2',3,3',6-Pentachlorobiphenyl

Chem. Abstr. Services Reg. No.: 52663-60-2
Chem. Abstr. Name: 2,2',3,3',6-Pentachloro-1,1'-biphenyl

2,2',3,4',6-Pentachlorobiphenyl

Chem. Abstr. Services Reg. No.: none available
Chem. Abstr. Name: 2,2',3,4',6-Pentachloro-1,1'-biphenyl

2,2',3,4,4',5'-Hexachlorobiphenyl

Chem. Abstr. Services Reg. No.: 35065-28-2
Chem. Abstr. Name: 2,2',3,4,4',5'-Hexachloro-1,1'-biphenyl

2,2',3,3',6,6'-Hexachlorobiphenyl

Chem. Abstr. Services Reg. No.: 38411-22-2
Chem. Abstr. Name: 2,2',3,3',6,6'-Hexachloro-1,1'-biphenyl

2,2',4,4',5,5'-Hexachlorobiphenyl

Chem. Abstr. Services Reg. No.: 35065-27-1
Chem. Abstr. Name: 2,2',4,4',5,5'-Hexachloro-1,1'-biphenyl

2,2',3',4,5,6'-Hexachlorobiphenyl

Chem. Abstr. Services Reg. No.: 38380-04-0
Chem. Abstr. Name: 2,2',3',4,5,6'-Hexachloro-1,1'-biphenyl

2,2',3,3',4,4',5-Heptachlorobiphenyl

Chem. Abstr. Services Reg. No.: 35065-30-6
Chem. Abstr. Name: 2,2',3,3',4,4',5-Heptachloro-1,1'-biphenyl

2,2',3,3',4,5,6'-Heptachlorobiphenyl

Chem. Abstr. Services Reg. No.: 38441-25-5
Chem. Abstr. Name: 2,2',3,3',4,5,6'-Heptachloro-1,1'-biphenyl

2,2',3,4,4',5,5'-Heptachlorobiphenyl

Chem. Abstr. Services Reg. No.: 35065-29-3
Chem. Abstr. Name: 2,2',3,4,4',5,5'-Heptachloro-1,1'-biphenyl

---

SOURCE: IARC (in press).

TABLE D.2 Composition of Chlorinated Biphenyls

Empirical Formula Chlorobiphenyls	Molecular Weight[a]	Percent Chlorine[a]	No. of Isomers
$C_{12}H_{10}$	154	0	1
$C_{12}H_9Cl$	188	18	3
$C_{12}H_8Cl_2$	222	31	12
$C_{12}H_7Cl_3$	256	41	24
$C_{12}H_6Cl_4$	290	48	42
$C_{12}H_5Cl_5$	324	54	46
$C_{12}H_4Cl_6$	358	58	42
$C_{12}H_3Cl_7$	392	62	24
$C_{12}H_2Cl_8$	426	65	12
$C_{12}HCl_9$	460	68	3
$C_{12}Cl_{10}$	494	79	1

[a]Based on Cl^{35}

SOURCE: Modified from Durfee et al. (1976).

al. 1976). The composition of some commercial mixtures from the United States, Italy, Germany, and Japan are illustrated in Table D.3 (IARC, in press; Durfee et al. 1976; Nisbet 1976).

Manufacturing Process

PCBs are produced by chlorination of biphenyl. The process, illustrated in Figure D.1, requires biphenyl, anhydrous chlorine, a catalyst such as iron filings or ferric chloride, and elevated temperatures. The crude product is purified by alkali wash, sometimes followed by distillation (Durfee 1976; IARC, in press; Nisbet 1976; Durfee et al. 1976). A detailed discussion of the manufacturing process of Monsanto (the U.S. producer) is given by Durfee.

The chemical composition of the final product varies with the degree of chlorination and with the blending process employed. Batches of PCBs with the same chlorine content may vary in composition and toxicity.

Formation of Unintentional Ingredients

There are no major compounds other than chlorobiphenyls found in commercially produced PCBs. The manufacturing process does not lead to the formation of unintentional ingredients measurable at 0.01 percent or more of product weight (Nisbet 1976). However, small amounts (see Table 3.2) of chlorodibenzofuran are detected in PCB mixtures, and quantities may vary from batch to batch.

The fact that different batches of PCBs, even from the same manufacturer, have different amounts of chlorodibenzofurans (CDFs) appears to be a reflection of variabilities in the production process. Whereas the primary process in the production of PCBs by the chlorination of biphenyl would not give CDFs, the aqueous alkaline washes and steam distillations could be anticipated to give the small amounts of CDFs observed, as shown in the following example.

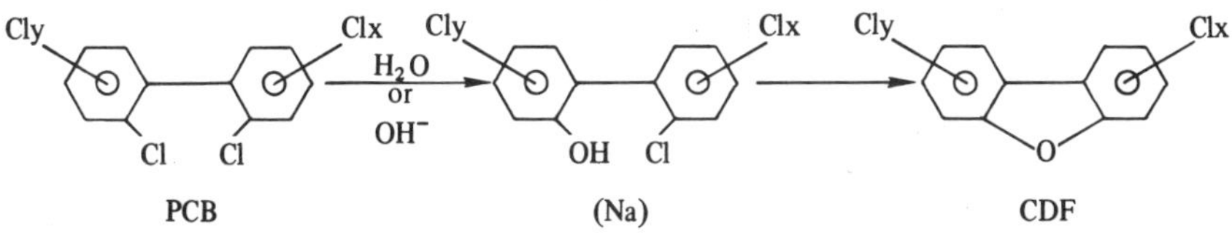

Such a scheme could have been involved in the Japanese "Kanemi Yusho" incident, where the PCB-contaminated rice oil was heated with steam at 200°, and where a relatively high ratio of CDFs to PCBs was observed, although Japanese investigators of the problem suggested an alternative route of contamination (Miyata et al. 1977).

PHYSICAL AND CHEMICAL DATA

These data have not been reported in a consistent manner in the scientific literature. Chemical and physical properties have been investigated for individual chlorobiphenyls in some instances and in others for the commercial product or mixture. An additional problem arises because two grades of PCBs existed. For nearly every Aroclor mentioned, a darker, less pure grade was available. The physical and chemical characteristics of the pure and technical grades were about the same, but

TABLE D.3 Approximate Percent Composition of Some Commercial PCB Products

Chlorobiphenyl	Aroclor Type or Grade 1016	1221	1232	1242	1248	1254	1260	Kanechlors KC-300	KC-400	KC-500	Fenclors DK	Clophen A60
$C_{12}H_{10}$	<0.1	11	6	<0.1	-	<0.1	-	-	-	-	-	-
$C_{12}H_9Cl$	1	51	26	1	-	<0.1	-	-	-	-	-	-
$C_{12}H_8Cl_2$	20	32	29	16	2	0.5	-	17	3	-	-	-
$C_{12}H_7Cl_3$	57	4	24	49	18	1	-	60	33	5	-	7
$C_{12}H_6Cl_4$	21	2	15	25	40	21	-	23	44	27	-	55
$C_{12}H_5Cl_5$	1	0.5	0.5	8	36	48	12	0.6	16	55	-	35
$C_{12}H_4Cl_6$	<0.1	-	-	1	4	23	38	-	5	13	-	3
$C_{12}H_3Cl_7$	-	-	-	<0.1	-	6	41	-	-	-	-	-
$C_{12}H_2Cl_8$	-	-	-	-	-	-	8	-	-	-	-	-
$C_{12}H_1Cl_9$	-	-	-	-	-	-	1	-	-	-	-	-
$C_{12}Cl_{10}$	-	-	-	-	-	-	-	-	-	-	100	-

SOURCE: IARC (in press), Nisbet (1976), Durfee et al. (1976).

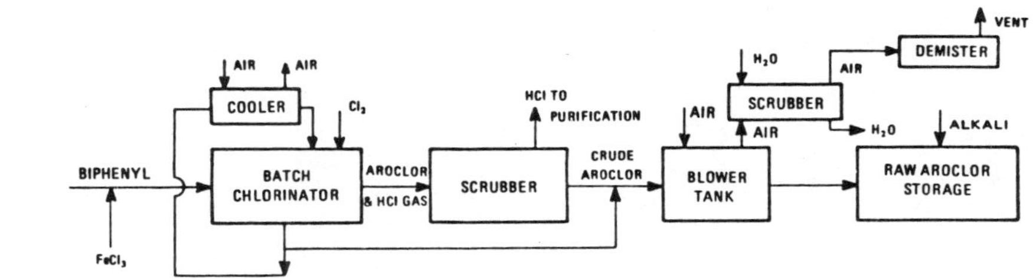

Preparation of crude chlorinated biphenyls

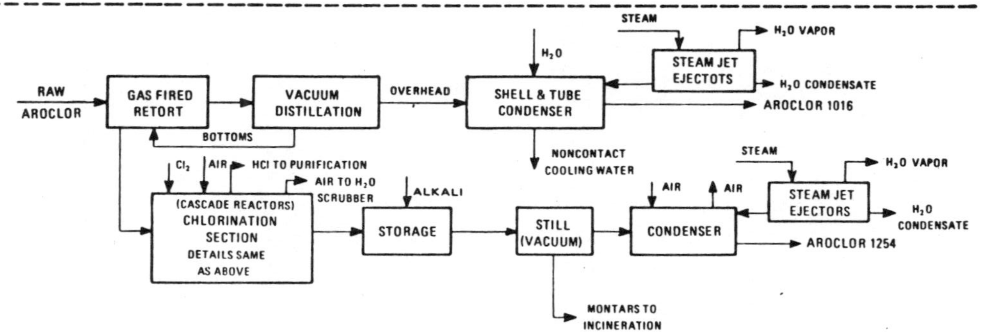

Distillation of crude products

SOURCE: Durfee (1976).

FIGURE D.1 Monsanto manufacturing process for PCBs.

the technical grades were lower in price (Monsanto Technical Bulletin 0/PL-306A). Impurities in lower grade Aroclors have not been determined, nor have these grades been characterized toxicologically. It is not known what percentage of total Aroclor usage in the United States was represented by the technical grade product. As far as we know, all toxicological studies have been done with the purified grades, and therefore data on technical grades are not available. Thus it is not possible to assess the toxicological effects that may be realized from exposure to technical grades of PCBs. (See Tables D.4-D.7.)

ENVIRONMENTAL CHEMISTRY DATA

Physical-Chemical Degradation

Hydrolysis PCBs may enter natural water through industrial discharge or as a result of disposal and cleanup of equipment. However, they are inert to hydrolysis except under extreme conditions (Hutzinger et al. 1974, Nisbet 1976).

Chemical Degradation PCBs are extremely stable compounds, not reactive chemically under environmental conditions. Oxidation, reduction, nitration, isomerization and nucleophilic reactions (reactions with alkali, alkoxides, amines) have been described (Hutzinger et al. 1974). The conditions required for these reactions make it unlikely that they will take place in the environment at any significant rate.

Some applications of PCBs, such as quenching of heated metals, heat-transfer media, and transformer oils, may lead to the formation of degradation products that can escape into the environment. The formation of chlorinated dibenzofurans on heating of PCBs to 500-600°C is one example (Buser et al. 1978). Polymeric products of unknown composition, possibly incorporating oxygen, are also formed from PCBs upon heating (Zitko and Choi 1971). Chemical contaminants (penta- to decachloroquaterphenyls, hepta- to nonachloroquaterphenyl ethers, and penta- to octachloroterphenyls), considered to be a result of dimerization were identified in Kanemi (Yusho) rice oils (Miyata et al. 1977). About 90 percent of this mixture consisted of the quaterphenyls, and their concentration equalled or exceeded, by 4 times, the concentration of PCBs. However, the analyses and identifications were carried out using saponified oil samples, and further confirmation of the results is required.

Photodegradation Photoreaction of PCBs produces a variety of products. Loss of chlorine, with replacement by hydrogen or hydroxyl groups, rearrangement, condensation, and "polar" products have been observed (Hutzinger et al. 1974.)

One of the reaction products of sunlight irradiation of PCB mixtures is the replacement of chlorine by hydroxy groups at the ortho positions. This allows oxygen to bind in a similar position on the other ring and to produce chlorodibenzofurans. Both heat and light can accelerate the transformation of PCBs to CDFs (Hutzinger et al. 1974).

It is not possible to assess quantitatively the importance of photodegradation of PCBs in the environment. A major portion of PCBs in the environment is not accessible to sunlight and, consequently, the amount of PCBs decomposed by photodegradation is likely to be very small. Table D.8 provides extinction coefficients and quantum yield data for the loss of a given PCB isomer (i.e., transformation by loss of a chlorine atom and formation of a PCB molecule with one less chlorine).

Biodegradation

Microbial degradation of PCBs depends on the degree of chlorination and the position of the chlorine atom on the biphenyl molecule. Lower chlorinated biphenyls are readily transformed by bacteria, but the higher chlorinated compounds are not (Nisbet 1976). *Achromobacter* isolated from sewage sludge rapidly degrades PCBs (Ahmed and Focht 1973). Kaiser and Wong (1974) also established microbial degradation using a bacterial culture isolated from lake water. Bacterial growth was not affected adversely when up to 0.1 percent Aroclor was added to the culture (Wong and Kaiser 1975). (See Table D.9.)

Biodegradation rates of chlorobiphenyls may be derived from the data of Furukawa et al. (1978b) using species of *Alcaligenes* and *Acinetobacter*. From these

TABLE D.4 Properties of Selected Aroclors

Property	1016	1221	1232	1242	1248	1254	1260
Color (APHA)	-	100	100	100	100	100	150
Physical state	mobile oil	mobile oil	mobile oil	mobile oil	mobile oil	viscous liquid	sticky resin
Stability	inert	inert	inert	inert	inert	inert	inert
Density (lb/gal 25°C)	11.40	9.85	10.55	11.50	12.04	12.82	13.50
Specific gravity x/15.5°C	1.36-1.37 x-25°	1.18-1.19 x-25°	1.27-1.28 x-25°	1.30-1.39 x-25°	1.40-1.41 x-65°	1.49-1.50 x-65°	1.55-1.56 x-90°
Distillation range (°C)	323-356	275-320	290-325	325-366	340-375	365-390	385-420
Acidity mg KOH/g, maximum	-	.014	.014	.015	.010	.010	.014
Fire point (°C)	none to boiling point	176	238	none to boiling point	none to boiling point	none to boiling point	none to boiling point
Viscosity (100°F)	71-81	38-41	44-51	82-92	185-240	1400-2500	-

SOURCE: Durfee et al. (1976).

TABLE D.5 Melting Points (°C) for Some PCB Isomers

Isomer	Melting Point (°C)
Monochlorobiphenyls	
2-	34
4-	77.7
Dichlorobiphenyls	
2,2'-	60.5
2,3'-	
2,4'-	44.5-46
4,4'-	149-150
Trichlorobiphenyls	
2,2',3-	28.1-28.8
2,3',4-	
2,4,4'-	57-58
2,2',5-	43-44
2,4',5-	67
2',3,4-	65-66
Tetrachlorobiphenyls	
2,3,4,4'-	142
2,2',3,5'-	49-50
2,2',4,5'-	66-68.5
2,3',4,4'-	127-127.5
2,2',5,5'-	87-89
2,3',4',5-	104-105
3,3',4,4'-	182-184
Pentachlorobiphenyls	
2,3,3',4,4'-	117-118.5
2,2',3,4,5'-	111.5-113
2,2',3,3',6-	
2,2',3,4',6-	
2,2',3,5',6-	98.5-100
2,3,3',4',6-	
2,2',3',4,5-	81-82
2,2',4,4',5-	
2,2',4,5,5'-	76.5-77.5
2,3',4,4',5-	112-113
Hexachlorobiphenyls	
2,2',3,4,4',5-	77-78
2,2',3,3',4,6-	
2,2',3,4,4',5'-	78.5-80
2,2',3,3',6,6'-	114-114.5
2,2',4,4',5,5'-	103-104
2,2',3',4,5,6'-	
Heptachlorobiphenyls	
2,2',3,3',4,4',5-	134.5-135.5
2,2',3,3',4,5,6'-	130.5-130.7
2,2',3,4,4',5,5'-	109-110

SOURCE: Modified from IARC (in press).

data, the mean primary biodegradation rates were calculated (Table D.10).

Considerable differences exist between the primary biodegradation rates of isomers (Furukawa et al. 1978b). Those substituted "in 2 and 6 positions" are degraded much more slowly than others, but the rates given in Table D.10 would probably be useful in designing a model.

According to Baughman and Lassiter (1978), representative concentrations of microbes in different water bodies are:

Water Body	Cell/ml
Pond	10^5
Stream	10^4
Eutrophic Lake	10^5
Oligotrophic Lake	10
Sediments	10^6

TABLE D.6 Solubility of Chlorobiphenyl Isomers and Aroclors in Water (mg/l)

Monochlorobiphenyls		Pentachlorobiphenyls	
2-	5.90	2,2',3,4,5'-	2.2×10^{-2}
3-	3.50	2,2',4,5,5'-	3.1×10^{-2}
4-	1.19		
		Hexachlorobiphenyl	
Dichlorobiphenyls			
		2,2',4,4',5,5'-	8.8×10^{-2}
2,4-	1.40		
2,2'-	1.50	Octachlorobiphenyl	
2,4'-	1.88		
4,4'-	0.08	2,2',3,3',4,4',5,5'-	0.7×10^{-2}
Trichlorobiphenyls		Decachlorobiphenyl	1.5×10^{-2}
2,4,4'-	8.5×10^{-2}		
2',3,4-	7.8×10^{-2}		
Tetrachlorobiphenyls			
2,2',5,5'-	4.6×10^{-2}		
2,2',3,3'-	3.4×10^{-2}		
2,2',3,5'-	17.0×10^{-2}		
2,2',4,4'-	6.8×10^{-2}	Aroclor 1242	0.24
2,3',4,4'-	5.8×10^{-2}	1248	5.40×10^{-2}
2,3',4',5-	4.1×10^{-2}	1254	1.20×10^{-2}
3,3',4,4'-	18.0×10^{-2}	1260	0.30×10^{-2}

SOURCE: Modified from Durfee et al. (1976).

Attempts to extrapolate the rates to environmental conditions introduce considerable uncertainty. The rates were measured under optimum conditions for the microorganism, at cell concentrations from 4.4×10^8 to 2×10^9 cells/ml and chlorobiphenyl concentrations of 50 nmol/ml, which is 11 mg/l to 15 mg/l for mono- to tetrachlorobiphenyls, respectively, well above the solubility of di- through tetrachlorobiphenyls (Weil et al. 1974).

If we assume that the rate depends linearly on the concentration of chlorobiphenyls, extrapolation to the "normal" ambient concentration in the ng/l range would involve multiplication by 10^{-9}. About 10 percent of the cells may be degrading PCBs (Wong and Kaiser 1975). Therefore, the rates should be multiplied by 10 percent of the normal cell concentration. The results of these computations are given in Table D.11.

Note that the biodegradation of most pentachlorobiphenyls may be extremely slow, and that of hexa- and more substituted chlorobiphenyls practically negligible. These rates are comparable to the rates of degradation by activated sludge illustrated in Figure D.2.

Effects of PCBs on microorganisms have been studied by many investigators and results are presented in Table D.12 (Bourquin and Cassidy 1975, Ewald et al. 1976, Nisbet 1976, Powers et al. 1977, Blakemore 1978, Blakemore and Carey 1978, Furukawa et al. 1978a).

PCBs may induce changes in the ecological role of microorganisms because of different sensitivities. Thus, community composition may change depending on level of PCBs, temperature, and the ability of the organisms to metabolize these compounds (O'Connors et al. 1978).

Mobility Through the Environment

Adsorption and Leaching Laboratory tests indicate that Aroclor 1254 is adsorbed most efficiently on soils with a high organic content and on clays. Aroclor 1016 was adsorbed strongly on silty clay loam soil and could be leached only with difficulty from silt and sandy loams. Less than 0.05 percent was leached after 4 months, and it consisted of mainly mono- and dichlorobiphenyls (Haque et al. 1974, Muto et al. 1974, Tucker et al. 1975a, Haque and Schmedding 1976, Nisbet 1976).

Transfer of PCB isomers from soil to water follows closely their physical properties, especially water solubility and partition coefficients as indicated in Table D.13 (Halter and Johnson 1977).

Volatility Monsanto Technical Bulletin 0/PL-306A gives a graph of vapor pressure (VP, mm Hg) of Aroclors 1242, 1248, 1254, and 1260 over the temperature range 150° to 300°C. From the graph, the logarithm of vapor pressure is a linear function of 1/T (T = temperature, °K): log(VP) = A-B/T

Aroclor	A	B	VP extrapolated to 20°C, mm Hg
1242	8.8	3,500	9.0×10^{-4}
1248	8.4	3,400	8.3×10^{-4}
1254	8.8	3,700	1.8×10^{-4}
1260	8.5	3,700	0.9×10^{-4}

TABLE D.7 Electrical Properties of Some Aroclors

Aroclor	Dielectric constant at 1000 cycles		Volume resistivity, 8 cm at 100°C 500 V, de	Dielectric strength,	Power factor 100°C, 1000 cycles, %
	25°C	100°C			
1232	5.7	4.6			
1242	5.8	4.9	above 500 x 10^9	>35	<0.1
1248	5.6	4.6	above 500 x 10^9	>35	<0.1
1254	5.0	4.3	above 500 x 10^9	>35	<0.1
1260	4.3	3.7	above 500 x 10^9	>35	<0.1
1268	2.5				

SOURCE: Modified from Durfee et al. (1976).

TABLE D.8 Photochemical Data for Selected PCBs in Aqueous Acetonitrile

Compound	Molecular Extinction Coefficients	Quantum Yield
4-Cl	120	0.0037
2,4-Cl_2	25	0.22
2,4,6-Cl_3	7	0.25
2,2',5,5'-Cl_4	30	0.017
Cl_{10}	1200	>0.21
Aroclor 1232	60	0.036
Aroclor 1268	770	0.16

SOURCE: Reprinted (with modifications) with permission from [Chemosphere 2:155-164, Bunce, N.J., Y. Kumar and B.G. Brownlee, An assessment of the impact of solar degradation of polychlorinated biphenyls in the aquatic environment], Copyright [1978], Pergamon Press, Ltd.

TABLE D.9 Microbial Species Used in Degradation Experiments

Species	Time	Degradation of PCB
Achrombacter		+ (unsubstituted ring)
Nocardia	20 days	+ (low chlorinated) <tetra CB
Aspergillus		-
Pseudomonces	73 days	+ (<tetra CB)
Acinetobacter		+

\+ indicates degradation.
\- indicates no effect.

TABLE D.10 Chlorobiphenyls Primary Biodegradation Rate

	nmol/cell hour
Mono	7×10^{-8}
Di-(both rings subst.)	6×10^{-8}
Tri-(both rings subst.)	5×10^{-8}
Tetra-(both rings subst.)	2.5×10^{-8}
Penta-*(both rings subst.)	1.5×10^{-8}

*Only one isomer, 2,4,5,2',5' studied.

TABLE D.11 Primary Biodegradation Rate (ng/ml-hr)

Chlorobiphenyls	Eutrophic Lake	Oligotrophic Lake	Sediment
Mono-	1.3×10^{-11}	1.3×10^{-15}	1.3×10^{-10}
Di-(both rings subst.)	1.3×10^{-11}	1.3×10^{-15}	1.3×10^{-10}
Tri-(both rings subst.)	1.3×10^{-11}	1.3×10^{-15}	1.3×10^{-10}
Tetra-(both rings subst.)	7.2×10^{-11}	7.3×10^{-17}	7.3×10^{-12}
Penta-(both rings subst.)	$<5 \times 10^{-14}$	$<5 \times 10^{-18}$	$<5 \times 10^{-13}$
More than penta no biodegradation			

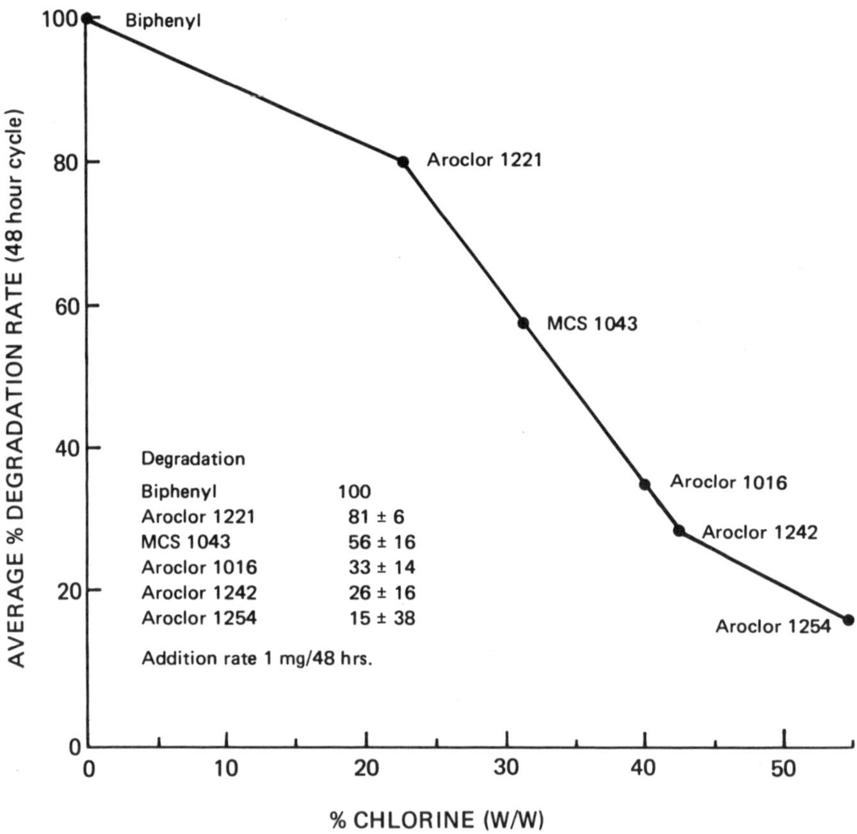

SOURCE: Tucker et al. (1975b).

FIGURE D.2 Semi-continuous activated sludge primary biodegradation rates of commercial PCBs as a function of the weight percent chlorine.

TABLE D.12 Effects of PCB on Selected Microorganisms

Species	Amount PCB	Effect
Aspergillus		inhibits growth
Acinetobacter		inhibits growth
Micrococcus/ Aerococcus		modified growth rate (reversible condition)
Scenedesmus	300-800μg/l	inhibit growth
Euglena	300-800μg/l	inhibit growth
Pseudomonad	10-100μg/l	reduced rate of cell division, reduced total amount and synthetic rate of nucleic acids
Cyclosterium	.1μg/l	reduced growth rate and chlorophyll levels
Thalassiosira	10μg/l	suppressed growth rate and photosynthesis

Dispersal through Water PCBs are strongly adsorbed to sediment and their dispersal through water depends on the movement of the sediment. (See discussion of the PCBs in Sediment in Chapter 1.)

Bioaccumulation

PCBs are remarkable among organic industrial chemicals for their low solubility in water (Table D.6), their high octanol/water partition coefficients, accumulation coefficients, (Tables D.14-D.16), and their resistance

TABLE D.13 Equilibrium PCB Concentrations and Distribution Coefficients (Kd) for Aroclor 1254

	Soil Concentration (mg/g)	H_2O Concentration (mg/l)	Mean (Kd)
Static Condition	500	7.60	5.6x 10^4
	350	7.10	
	200	3.90	
	100	1.70	
Continuous Flow	500	.54	93 x10^4
	350	.37	
	200	.21	
	100	.11	

SOURCE: Modified from Halter and Johnson (1977).

TABLE D.14 Octanol/Water Partition Coefficients and Ecological Magnification Values for Selected PCBs

	Octanol/Water Partition Coefficient	Ecological Magnification Value
4-chloro	390	480
4,4'-dichloro	5,700	1,200
2,5,2'-trichloro	7,800	6,400
2,5,2',5'-pentachloro	8,100	12,000
2,4,5,2',4',5'-hexachloro	17,000	12,000
decachloro	30,000	42,000
	190,000	97,000

to in vivo degradation (Table D.17). As a result, they exhibit extraordinarily high values for bioaccumulation in animal tissues, especially in fish and other aquatic organisms. For example, the average concentration of PCBs in Lake Michigan is about 0.008 μg/kg, yet mature lake trout, Salvelinus namaycush, contain an average of 28 mg/kg, equivalent to a bioconcentration factor of 3.41×10^6 (Metcalf 1977, U.S. EPA 1972).

Bioaccumulation of PCBs by fish and other aquatic organisms may take place either through ingestion of contaminated food organisms or by direct absorption through the integument. Coho salmon, Oncorhynchus kisutch, were fed food pellets contaminated with 3,4,3',4'-tetrachlorobiphenyl, 2,4,6,2',4',6'- and 2,4,5,2',4',5'-hexachlorobiphenyl. After 108 days, the salmon contained an average of 1.4 mg/kg of each of the two hexachlorobiphenyls and 0.65 mg/kg of tetrachlorobiphenyl. From 30 to 60 percent of the

TABLE D.15 PCB Accumulation Coefficients in Selected Animals

Animal	Accumulation Coefficients
oysters (Aroclor 1254)[1]	101×10^3 - 165×10^3
shrimp (Aroclor 1254)[1]	26×10^3
fish (estuarine) (Aroclor 1254)[1]	37×10^3
zooplankton (dichloro-)[2]	3 - 100 (% bdy wt $\times 10^{-4}$)
trout (tetrachloro-)[3]	9×10^3

SOURCE:
[1]Modified from Hansen (1976).
[2]Modified from McNaught (1978).
[3]Modified from Branson et al. (1975).

TABLE D.16 Accumulation Coefficient of Aroclor 1254 by Aquatic Invertebrates

Organism	Organisms Per Sample[a]	Water Concentration[b] (μg/kg)	Organism Concentration[b] (4-day exposure) (mg/kg)	Accumulation Factor[c] (x 10^3)				
				1-day	4-day	7-day	14-day	21-day
Daphnid *Daphnia magna*	60	1.1 ± 0.2	52.0 ± 2.0	25.0	47.0	---	---	---
Phantom midge *Chaoborus punctipennis*	5	1.3 ± 0.1	30.0 ± 1.6	22.0	23.0	24.0	25.0	---
Scud *Cammarus pseudolimnaeus*	6	1.6 ± 0.1	39.0 ± 3.0	17.0	24.0	26.0	28.0	27.0
Mosquito larvae *Culex tarsalis*	10	1.5 ± 0.3	27.0 ± 2.0	13.0	18.0	20.0	---	---
Glass shrimp *Palaemonetes kadiakensis*	3	1.3 ± 0.1	16.0 ± 2.6	10.0	12.0	14.0	14.0	17.0
Stonefly *Pteronarcys dorsata*	3	2.8 ± 0.8	7.0 ± 0.3	2.1	2.5	2.8	2.9	2.8
Dobsonfly *Carydalus cornutus*	3	1.1 ± 0.1	5.1 ± 0.2	1.4	4.6	5.7	6.6	6.8
Crayfish *Orconectes nais*	2	1.2 ± 0.1	0.2 ± 0.2	0.57	1.7	3.4	4.5	5.1

[a]Samples were taken in triplicate.
[b]Samples were taken in triplicate and expressed as mean value ± standard error (p = .05).
[c]Concentration in organism/concentration in water.

SOURCE: Modified from Nisbet (1976).

TABLE D.17 Ecological Magnification (EM) and Biodegradability Index (BI) Compared with Water Solubility and Partition Coefficients

Chemical	Water Solubility (ppb)	Partition Coefficient octanol/water	EM ($x 10^3$)				BI ($x 10^{-2}$)			
			Alga	Snail	Mosquito	Fish	Alga	Snail	Mosquito	Fish
tri-Cl-PCB (2,5,2')	16	7,800	7.3	5.8	.8	6.4	30.0	17.0	35.0	60.0
tetra-Cl-PCB (2,5,2',5')	16	8,100	18.0	39.0	10.6	12.0	1.5	8.2	7.6	6.0
penta-Cl-PCB (2,4,5,2',5')	19	16,000	5.5	60.0	17.0	12.0	2.9	2.7	1.3	1.9
hexachloro (2,4,5,2',4',5')	9	30,000	9.4	101.0	105.0	42.0	0.8	2.1	0.7	0.8
decachloro		190,000	100.0	930.0	22.0	97.0	---	---	---	0.5

chlorobiphenyls in the diet was retained in the fish after termination of feeding of the chlorobiphenyls. The level of the tetrachlorobiphenyl in the adipose tissue decreased from 10.8 mg/kg to 2.5 mg/kg over a period of 48 days, compared with 2,4,5,2',4',5'-hexachlorobiphenyl, which decreased from 20.1 mg/kg to 6.5 mg/kg; and with 2,4,6,2',4',6'-hexachlorobiphenyl, which decreased from 20.2 mg/kg to 6.0 mg/kg (Gruger et al. 1975).

The direct uptake of ^{14}C-2,5,2'-trichloro-, 2,5,2',5'-tetrachloro-, and 2,4,5,2',5'-pentachlorobiphenyls by the green sunfish, _Lepomis cyanellus_, was studied by Sanborn et al. (1975). After the fish had spent 15 days in static water, the bioconcentration factors were determined as trichloro- 54, tetrachloro- 460, and pentachlorobiphenyl 1,510.

Model ecosystem studies Bioaccumulation and biodegradation studies with a series of ^{14}C-labeled chlorobiphenyls have been made using the terrestrial aquatic model ecosystem involving both food-chain transfer and direct absorption in aquatic organisms, including the snail, _Physa_ sp., and the fish _Gambusia affinis_ (Metcalf et al. 1975, Metcalf and Lu 1978). The degree of bioaccumulation, as determined from the ratio of amount of parent compound in snail or fish to amount in water, increased in proportion to the degree of chlorination; and it reached high values for hexachlorobiphenyl--42 x 10^3 in fish and 100 x 10^3 in snail--and for decachlorobiphenyl--97 x 10^3 in fish and 930 x 10^3 in snail.

Correspondingly, the higher chlorobiphenyls were more stable in the tissues of the organisms, as determined by biodegradability index or ratio of amount of polar compound stored. The biodegradability index in the fish decreased from 0.60 for trichlorobiphenyl to 0.005 for decachlorobiphenyl.

HAZARD EVALUATION

Acute Toxicity Tests

Lethal Dose Studies LD_{50} and LC_{50} for aquatic and terrestrial wildlife are presented in Tables D.18-D.22. Oral doses range from 4,000 mg/kg to 11,000 mg/kg, and this level is required to obtain LD_{50}s depending upon

TABLE D.18 Toxicity of PCBs and DDT in Continuous Flow Tests

		LC_{50} (μg/l)					
Aroclor	Species	5 days	10 days	15 days	20 days	25 days	30 days
1254	Rainbow Trout	150.0	8.0	----	----	----	----
1260		----	240.0	94.0	21.0	----	----
DDT		2.3	0.9	0.3	----	----	----
1242	Bluegills	150.0	72.0	54.0	----	----	----
1248		310.0	160.0	76.0	10.0	----	----
1254		----	440.0	200.0	140.0	54.0	----
1260		----	----	----	240.0	210.0	150.0
1242	Channel Catfish	----	170.0	110.0	----	----	----
1248		----	230.0	130.0	----	----	----
1254		----	----	740.0	300.0	110.0	----
1260		----	----	----	300.0	170.0	140.0
1248[a]	Bluegills	140.0	76.0	----	----	----	----
1248[a]	Channel Catfish	----	94.0	57.0	----	----	----

[a]Temperature, 27°C.

SOURCE: Stalling and Mayer (1972).

TABLE D.19 Toxicity of PCBs and DDT to Invertebrates

Compound	Organism	Bioassay Type[a]	Exposure (days)	LC_{50} (μg/l)
Aroclor 1242	Crayfish	static	7	30
Aroclor 1254	Crayfish	static	7	100
Aroclor 1254	Crayfish	continuous-flow	7	80
DDT	Crayfish	static	4	100
Aroclor 1242	Scud	continuous-flow	4	10
Aroclor 1242	Scud	continuous-flow	10	5.0
Aroclor 1248	Scud	static	4	52
Aroclor 1254	Scud	static	4	2,400
DDT	Scud	static	4	3.2
DDT	Scud	continuous-flow	5	0.6
Aroclor 1254	Glass Shrimp	continuous-flow	7	3.0
DDT	Glass Shrimp	static	5	1.0
DDT	Glass Shrimp	continuous-flow	5	1.3
Aroclor 1242	Dragonfly	static	7	800
Aroclor 1254	Dragonfly	static	7	1,000
Aroclor 1242	Damselfly	continuous-flow	4	400
Aroclor 1254	Damselfly	continuous-flow	4	200
DDT	Damselfly	static	4	56

[a]Temperature, 15.6°C; alkalinity, 35ppm; pH, 7.1.

SOURCE: Stalling and Mayer (1972).

TABLE D.20 Toxicity of PCBs to Rats and Rabbits, LD_{50} (mg/kg of body weight)

	Aroclor 1221	Aroclor 1232	Aroclor 1242	Aroclor 1248	Aroclor 1254	Aroclor 1260	Aroclor 1268
Rat (oral)	4000	4500	8700	11,000	----	10,000	11,300
Adult rats (oral)					4000-10,000	4000-10,000	
Weaning rats (oral)					1200	1300	
Adult rats (i.v.)					400		
30-day old rats (oral)					1400		
120-day old rats (oral)					2000-2500		
Rabbit (skin)	>2000 <3200	>1300 <2000	>800 <1300	>800 <1300		>1300 <2000	<2500

SOURCE: Modified from IARC (in press).

TABLE D.21 Acute Oral and Dermal Toxicity Effects of PCB Exposure

Preparation	Animal	Treatment	Percent Mortality	Liver Effects	Skin Effects
<u>Oral</u>					
Unknown	Mouse	Single dose 2000mg/kg			
54%Cl	Rat	Single dose 500mg/kg	0	Increase of weight and lipid; potentiation of CCl_4 toxicity	
42,54,60, and 68%Cl	Mallard	Single dose 2000mg/kg	0		
42%Cl	Rat	20 daily doses 138mg	0 in 3 months	Hyalin bodies in liver cells	
42%Cl	Guinea Pig	2 doses (1 week apart) 69mg	100 between 11 and 29 days	Fatty metamorphosis; central atrophy	
65%Cl	Rat	6 daily doses 300mg	70 in 14 days	Increase of weight; cell swelling; hyalin granules	
65%Cl	Rat	50mg every second day	60 in 5 weeks	33% weight increase; cell swelling; hyalin granules	
<u>Dermal</u>					
42%Cl	Guinea Pig	11 daily, 34.5mg	100 between 11 and 21 days	Fat; central atrophy perinuclear basophilic granulation; focal necrosis in a few animals	Occasional thickening of the epidermis
42%Cl	Rabbit	Alternate days; total dose from 946 to 1980mg	100 between 17 and 98 days	Fatty degeneration; central atrophy	Thinning of prickle cell layer and thickening of outer cornified layers
Aroclor	Rabbit	Daily, 0.3, 0.6, and 0.9g	High dose died before liver necrosis developed	Moderate doses; mottled liver, subacute yellow atrophy; fatty degeneration and marked necrosis	Reddening; formation of small papules and blisters; finally desquamation of external epidermal layers

SOURCE: Modified from Nisbet (1976).

TABLE D.22 LC_{50} Comparisons of Aroclors in Diets (mg PCB/kg Diet) of 2-Week-Old Birds

	Mallard	Pheasant	Bobwhite Quail	Japanese Quail
Aroclor 1232	----	3200	3000	>5000
Aroclor 1242	3200	2100	2100	>5000
Aroclor 1248	2800	1300	1200	4800
Aroclor 1254	2700	1100	600	2900
Aroclor 1260	2000	1100	750	2200
Aroclor 1262	3000	1200	870	2300

SOURCE: Modified from Nisbet (1976).

the Aroclor mixture. PCBs are virtually nontoxic to rats, but acutely toxic to some invertebrates (LC_{50} generally less than 1 mg/l).

Life-cycle Data Tests using Daphnia, fathead minnow, Jordanella, and Gammrus indicated that the larval stage is most sensitive to PCB effects (Nebeker et al. 1977). (See Tables D.19, D.23, and D.24.) Life-cycle effects on birds are given in Table D.25.

Metabolic and Chronic Feeding Studies The predominant route of PCB metabolism is hydroxylation (Table D.26). Some chlorobiphenyls may be hydroxylated via arene oxide intermediates which suggests potential carcinogenicity and mutagenicity. The hydroxylated chlorobiphenyls are more toxic than the parent compounds. The toxicological significance of the recently discovered chlorobiphenyl methyl sulfones is not known. Results of chronic feeding studies are presented in Table D.27.

Mutagenicity For Aroclors 1242 and 1254 no statistical evidence has been presented for mutagenic effects in bone marrow or spermatogonial cells, for mutagenic effects in dominant lethal assays, or for chromosomal aberrations in human lymphochromosomal cultures. Ames test results with Aroclor 1221 and 4-chlorobiphenyl indicated significant mutagenicity (Wyndham et al. 1976).

Oncogenicity Tests Data indicate that Kaneclor 500, Aroclor 1260, and perhaps Aroclor 1254 are carcinogenic in mice and rats (Table D.27).

TABLE D.23 Spawning and Egg Production of Fathead Minnows Exposed to PCBs

Aroclor concentration (μg/l)	Spawnings/ female	Eggs/ spawning	Eggs/ female	Percent hatched
Aroclor 1242				
51.0	0	0	0	0
15.0	0	0	0	0
5.4	2.5	30	150	81
2.9	3.9	63	280	38
0.9	1.3	28	35	84
0.0	4.6	90	440	53
Aroclor 1254				
15.0	0	0	0	0
4.6	0	0	0	0
1.8	1.0	63	110	79
0.5	5.2	100	560	63
0.2	3.0	64	220	55
0.0	2.4	100	250	75

SOURCE: Modified from Nisbet (1976).

TABLE D.24 Incubation Times and Hatchabilities of Coho Salmon Eggs, and Survival and Sizes of Alevins 4 Weeks After Hatching, After Exposure to Aroclor 1254 for 6 and 2 Weeks at 12-14°C

	Incubation			Mean Alevin Size	
PCB (μg/kg)	Days	Percent Hatchability	Percent Survival	Length (mm)	Weight (g)
Exposure for 6 weeks (2 before and 4 after hatching)					
0	510	96	91	36	0.30
5	490	88	76	35	0.30
10	490	78	64	34	0.30
15	---	38	24	32	0.27
25	460	96	77	30	0.28
55	460	63	7	28	0.24
Exposure for 2 weeks (to 2 days before hatching)					
0	500	93	93	35	0.30
5	480	96	70	35	0.30
10	490	90	88	35	0.30
15	490	92	84	33	0.26
25	480	98	75	35	0.29
55	490	66	40	34	0.26

SOURCE: Modified from Nisbet (1976).

TABLE D.25 Life Cycle Effects of Aroclor 1254

Species	Dose (mg/kg)	Effects
Pheasants	50 and 200	reduced egg production; increase in no. of eggs pipped but not hatched; reduced survival of chicks and overall reproductive success
Mallard*	25	no reproductive effects; PCB residue in eggs equalled 33 ppm wet weight
Bobwhite Quail*	50	no reproductive effects
Ring Doves	10	high embryonic mortality in 2nd generation; chromosomal aberrations; abnormal incubation behavior seen in parents
Am. Kestrels	10	slight increase in egg shell thickness
Japanese Quail	78	slight reduction in egg production; 12% reduction in egg shell thickness
	50	non significant reproductive effects

*Required test species in proposed guidelines.

TABLE D.26 Metabolism of Polychlorinated Biphenyl Isomers

Species	Parent Compound	Products
rat	4-chlorobiphenyl	monohydroxylated derivative dihydroxylated derivative
rat	4-chlorobiphenyl	4'-chloro-3,4-biphenyldiol 4'-chloro-3-methoxy-4-biphenyldiol 4'-chloro-4-methoxy-3-biphenyldiol 4'-chloro-4-methoxy-3,5-biphenyldiol
rat	^{14}C-2,3 dichlorobiphenyl	2',5'-dichloro-4-biphenyl
rat	^{14}C-2,4,6-trichlorobiphenyl	4'-hydroxylated derivative 3',4'-dihydroxylated derivative 3'-hydroxy-4'-methoxylated derivative 4'-hydroxy-3'-methoxylated derivative
rat microsomes	2,2',4-trichlorobiphenyl	monohydroxylated derivative
rat	^{3}H-2,2',5,5'-tetrachlorobiphenyl	3-monohydroxylated derivative
rat	^{3}H-2,2',5,5'-tetrachlorobiphenyl	2,2',5,5'-tetrachloro-3-biphenylol *trans*-3,4-dihydro-2,2',5,5'-tetrachlorobiphenyl-3,4-biphenylol
rat	3,3',4,4'-tetrachlorobiphenyl	2 and 5-hydroxylated derivative 3 and 5-hydroxylated derivative
rat	^{14}C-2,3,4,5,6-pentachlorobiphenyl	3' and 4'-hydroxylated derivative 3',4'-dihydroxylated derivative
rat	2,2',4,5,5'-pentachlorobiphenyl	2,2',4,5,5'-pentachloro-3'-biphenylol 3',4'-dihydro-3,4'-diol 2,2',4,5,5'-pentachloro-3,4-biphenylol
rat	2,2',4,4',6-pentachlorobiphenyl 2,2',4,6,6'-pentachlorobiphenly 2,3,4,4',6-pentachlorobiphenyl 2,3',4,5',6-pentachlorobiphenly	monohydroxylated derivative monohydroxylated derivative monohydroxylated derivative monohydroxylated derivative
rat	2,3,3',4,4'-pentachlorobiphenyl	no hydroxylated products

TABLE D.26 continued

Species	Parent Compound	Products
rat	2,2',4,4',5,5'-hexachlorobiphenyl	no hydroxylated products
rat	^{14}C-2,4,6,2',4',6'-hexachlorobiphenyl	monohydroxylated derivative
rat	2,2',4,4',5,5'-hexachlorobiphenyl	2,2',4,4',5,5'-hexachloro-3-biphenylol
rabbit	4-chlorobiphenyl	4-chloro-4'-biphenylol
rabbit (liver microsomes)	4-chlorobiphenyl	4-chloro-4'-biphenylol 4'-chloro-3,4-biphenyldiol
rabbit	4,4'-dichlorobiphenyl	4,4'-dichloro-3-biphenylol 3,4'-dichloro-4-biphenylol 4'- chloro-4-biphenylol
rabbit	2,2',5,5'-tetrachlorobiphenyl	2,2',5,5'-tetrachloro-3-biphenylol 2,2',5,5'-tetrachloro-4-biphenylol _trans_-3,4-dihydro-2,2',5,5'-tetrachloro-3,4-biphenyldiol
rabbit	2,2',4,4',5,5'-hexachlorobiphenyl	monohydroxylated derivative monohydroxy with chlorine shift monohydroxy methoxylated derivative
pigeons	4-chlorobiphenyl	monohydroxylated derivative
pigeons	4,4'-dichlorobiphenly	monohydroxylated derivative
pigeons	2,2',5,5'-tetrachlorobiphenyl	monohydroxylated derivative

pigeons	2,2',4,4',5,5'-hexachlorobiphenyl	no hydroxylated products
chickens	2,2',4,4',5,5'-hexachlorobiphenyl	metahydroxylated derivative pentachlorobiphenyl derivative pentachloro trihydroxylated derivative
trout	4-chlorobiphenyl	no hydroxylated products
trout	4,4'-dichlorobiphenyl	no hydroxylated products
trout	2,2',5,5'-tetrachlorobiphenyl	no hydroxylated products
trout	2,2',5,5'-tetrachlorobiphenyl	conjugated metabolites 4-hydroxy-2,2',5,5'-TCB
trout	2,2',4,4',5,5'-hexachlorobiphenyl	no hydroxylated metabolites
goat	4-chlorobiphenyl	4'-chloro-4-biphenylol 4'-chloro-3,4-biphenyldiol
goat	4,4'-dichlorobiphenly	4,4'-dichloro-3-biphenylol
cow	4-chlorobiphenyl	4'-chloro-4-biphenylol
rhesus monkey	2,2',5,5'-tetrachlorobiphenly	monohydroxylated derivatives dihydroxylated derivatives *trans*-3,4,dihydro-3,4-TCB diol hydroxy-3,4-dihydro-3,4-TCB diol
mice	2,2',5,5'-tetrachlorobiphenly	methylsulfone derivative
mice	pentachlorobiphenly	methylsulfone derivative
mice	2,2',4,4',5,5'-hexachlorobiphenyl	2,2',4,4',5,5'-hexachloro-3-biphenylol

SOURCE: Modified from IARC (in press).

TABLE D.27 Results of Chronic Feeding Studies Using Various PCB Mixtures

Species	Dose	PCB	Time	Effect
Mouse	500μg/g or less	Kanechlor 500	32 weeks	Hepatocellular carcinomas; some wide-spread nodular hyperplasia (Nagasaki et al. 1972, Ito et al. 1973)
	500μg/g	Kanechlor 300, 400	32 weeks	No effect (Nagasaki et al. 1972)
	300μg/g	Aroclor 1254	11 months	Adenofibrosis of liver (Kimbrough and Linder 1974)
	300μg/g	Aroclor 1254	6 months	No effect (Kimbrough 1974)
Rat	100 & 500 μg/g	Aroclor 1254	8 months	Adenofibrosis (Kimbrough et al. 1972)
	38 - 616 μg/g	Kanechlor 400	57 weeks	Females developed neoplastic nodules in liver (Kimura and Baba 1973)
	100μg/g	Aroclor 1260	21-23 months	Increased number of stillborns and reduction of fetal weight; neoplastic nodules of liver (Burke and Fitzhugh 1970)
	100μg/g	Aroclor 1254 & 1260	1 year	No illness; increase in total serum lipids & cholesterol; slight increase in triglycerides; liver hypertrophy and focal areas of hepatocellular degeneration (IARC [in press])
	20 - 100 μg/g	Kanechlor 500	2 weeks (gestation)	Offspring with reduced learning ability (Shiota 1976)
Monkey	2.5 - 5.0 μg/g	1248	6 months	Females developed acne, alopecia, erythema and swelling of eyelids; alteration of menstrual cycles; increased frequency of abortion (IARC [in press])

			14 months	Males developed only slight periorbital oedema and erythema (Barsotti et al. 1976)
				Hypertrophia hyperplastic gastritis, ulceration and anemia (IARC [in press])
			2 months	Infants with evidence of severe hyperplastic gastritis and hyperplasia of hair follicle epithelium and epidermis (IARC [in press])
Chickens	50µg/g	1254	39 weeks	Reduced egg production and decline in fertility (Platonou and Reinhart 1973)
	50mg/l in H_2O	1254	6 weeks	Deformities in neck and toe (Bush et al. 1974)
	100µg/g	1254		Decrease hatchability (Keplinger et al. 1971)
	20µg/g		8 weeks	Subcutaneous edema; atrophy of the spleen; mild necrosis; congestion and fatty infiltration of the liver 30% mortality (Bird et al. 1978)
Mink	10µg/g	1254	4 months	Dose response relationship in depressed weight gain; reduced reproduction (Aulerich and Ringer 1977)
			5.5 months	Mortality occurred (Aulerich and Ringer 1977)
Pelican	100mg		10 weeks	Increased size of hepatocytes; number of vacuoles per hepatocyte and number of mitochondria (Stotz and Greichus 1978)
Dogs	100µg/g	1260	12 months	Males - reduced growth rate; increased liver weights; elevated serum alkaline phosphate (Keplinger et al. 1971)

REFERENCES

Ahmed, M. and D.D. Focht (1973) Degradation of polychlorinated biphenyls by two species of Arochromobacter. Can. J. Microbiol. 19:47-52.

Aulerich, R.J. and R.K. Ringer (1977) Current status of PCB toxicity to mink, and effect on their reproduction. Arch. Environ. Contam. Toxicol. 6:279-292.

Barsotti, D.A., R.J. Marlar, and J.R. Allen (1976) Reproductive dysfunction in rhesus monkeys exposed to low levels of polychlorinated biphenyls (Aroclor 1248). Food Cosmet. Toxicol. 14:99-103.

Baughman, G.L. and R.R. Lassiter (1978) Prediction of environmental pollutant concentration. Pages 35-54, Estimating the Hazard of Chemical Substances to Aquatic Life. Edited by John Cairns, Jr., K.L. Dickson, and A.W. Maki. Philadelphia: American Society for Testing and Materials.

Bird, F.H., C.B. Chawan, and R.W. Gerry (1978) Response of broiler chickens to low level intake of polychlorinated biphenyl isomers. Poultry Sci. 57(2):538-541.

Blakemore, R.P. (1978) Effects of polychlorinated biphenyls on macromolecular synthesis by heterotrophic marine bacterium. Appl. Environ. Microbiol. 35(2):329-336.

Blakemore, R.P. and A.E. Carey (1978) Effects of polychlorinated biphenyls on growth and respiration of heterotrophic marine bacteria. Appl. Environ. Microbiol. 35(2):323-328.

Bourquin, A.W. and S. Cassidy (1975) Effect of polychlorinated biphenyl formulations on the growth of estuarine bacteria. Appl. Microbiol. 29(1):125-127.

Branson, D.R., G.E. Blau, H.C. Alexander, and W.B. Neely (1975) Bioconcentration of 2,2',4,4'-tetrachlorobiphenyl in rainbow trout as measured by an accelerated test. Trans. Am. Fish Soc. 4:785-792.

Bunce, N.J., Y. Kumar, and B.G. Brownlee (1978) An assessment of the impact of solar degradation of polychlorinated biphenyls in the aquatic environment. Chemosphere 2:155-164.

Burke, J. and O.G. Fitzhugh (1970) Status Report of Chemistry and Toxicology of PCBs, Suppl. I, December 1. Washington, D.C.: U.S. Food and Drug Administration.

Buser, H.R., H.P. Boffhardt, and C. Rappe (1978) Formation of polychlorinated dibenzofurans (PCDFs) from the pyrolysis of PCBs. Chemsphere 7:109.

Bush, B., C.F. Tumasonis, and F.D. Baker (1974) Toxicity and persistence of PCB hemologs and isomers in the avian system. Arch. Environ. Contam. Toxicol. 2:195-212.

Durfee, R.L. (1976) Production and usage of PCBs in the United States. Pages 103-107, Proceedings of National Conference on Polychlorinated Biphenyls, November 19-21, 1975, Chicago, Illinois. Report No. EPA-560/6-75-004. Washington, D.C.: U.S. Environmental Protection Agency.

Durfee, R.L., G. Contos, F.C. Whitmore, J.D. Barden, E.E. Hackman III, and R.A. Westin (1976) PCBs in the United States--Industrial Use and Environmental Distribution. Report No. EPA 56016-16-005. Springfield, Va.: Versar, Inc.

Ewald, W.G., J.E. French, and M.A. Champ (1976) Toxicity of polychlorinated biphenyls (PCBs) to <u>Euglena gracilis</u>: Cell population growth, carbon fixation, chlorphyll level, oxygen consumption, and protein and nucleic acid synthesis. Bull. Environ. Contam. Toxicol. 16(1):71-80.

Furukawa, K., F. Matsumura, and K. Tonomura (1978a) <u>Alcaligenes</u> and <u>Acinetobacter</u> strains capable of degrading polychlorinated biphenyls. Agric. Biol. Chem. 42(3):543-548.

Furukawa, K., K. Tonomura, and A. Kamibayashi (1978b) Effect of chlorine substitution on the biodegradability of polychlorinated biphenyls. Appl. Environ. Microbiol. 35(2):223-227.

Gruger, H.E., N.L. Karrich, A.I. Davidson, and T. Hruby (1975) Accumulation of 3,4,3',4'-tetrachlorobiphenyl and 2,4,5,2',4',5'- and 2,4,6,2',4',6'-hexachlorobiphenyl in juvenile Coho salmon. Environ. Sci. Technol. 9:121-127.

Halter, M.T. and H.E. Johnson (1977) A model system to study the desorption and biological availability of PCB in hydrosoils. Aquat. Toxicol. Haz. Eval. ASTM-STP 634:178-195.

Hansen, D.J. (1976) PCBs: Effect on and accumulation by estuarine organisms. Pages 282-283, Proceedings of National Conference on Polychlorinated Biphenyls, November 19-21, 1975, Chicago, Illinois. Report No. EPA-560/6-75-004. Washington, D.C.: U.S. Environmental Protection Agency.

Haque, R. and D. Schmedding (1976) Studies on the adsorption of selected polychlorinated biphenyl isomers on several surfaces. J. Environ. Sci. Health B11(2):129-137.

Haque, R., D.W. Schmedding, and V.H. Freed (1974) Aqueous solubility, adsorption, and vapor behavior of polychlorinated biphenyl Aroclor 1254. Environ. Sci. Technol. 8(2):139-142.

Hutzinger, O., S. Safe, and V. Zitko (1974) The Chemistry of PCBs. Cleveland, Ohio: CRC Press.

International Agency for Research on Cancer (IARC) (In press) Monographs on the evaluation of the carcinogenic risk of chemicals to humans 18:43-103. (Due for release in or after 1979)

Ito, N., H. Nagasaki, M. Arai, S. Makiura, S. Sugihara, and K. Hirao (1973) Histopathologic studies on liver tumorigenesis induced in mice by technical polychlorinated biphenyls and its promoting effect on liver tumors induced by benzene hexachloride. Gann 51:1637-1646.

Kaiser, K.L.E. and P.T.S. Wong (1974) Bacterial degradation of polychlorinated biphenyls. I. Identification of some metabolic products from Aroclor 1242. Bull. Environ. Contam. Toxicol. 11(3):291-296.

Keplinger, M.L., O.E. Fancher, and J.C. Calandra (1971) Toxicologic studies with polychlorinated biphenyls. Toxicol. Appl. Pharmacol. 19:402-403.

Kimbrough, R.D. (1974) The toxicity of polychlorinated polycyclic compounds and related chemicals. In, CRC Crit. Rev. Toxicol. Pages 445-498, Cleveland, Ohio: CRC Press.

Kimbrough, R.D. and R.E. Linder (1974) Induction of adenofibrosis and hepatomas of the liver in BALB/cJ mice by polychlorinated biphenyls (Aroclor 1254). J. Nat. Cancer Inst. 53:547-552.

Kimbrough, R.D., R.E. Linder, and T.B. Gaines (1972) Morphological changes in livers of rats fed polychlorinated biphenyls. Arch. Environ. Health 25:354-364.

Kimura, N.T. and T. Baba (1973) Neoplastic changes in the rat liver induced by polychlorinated biphenyl. Gann 64:105-108.

McNaught, D.C. (1978) Uptake Rates and Effects of PCBs on Phytoplankton and Zooplankton. (Unpublished paper)

Metcalf, R.L. (1977) The biological fate and transformation of pollutants in water. Pages 195-221, Fate of Pollutants in the Air and Water Environment, Part II. Edited by I.M. Suffet. New York: John Wiley & Sons, Inc.

Metcalf, R.L. and Po-Yung Lu (1978) Partition Coefficients as Measures of Bioaccumulation Potentials for Organic Compounds. Final Report to U.S. Environmental Protection Agency, Contract No. 68-014189, July 15. Urbana, Ill.: University of Illinois.

Metcalf, R.L., J.R. Sanborn, Po-Yung Lu, and D. Nye (1975) Laboratory model ecosystem studies of the degradation and fate of radiolabeled tri-, tetra-, and pentachlorobiphenyl compared with DDE. Arch. Environ. Contam. Toxicol. 3:151-165.

Miyata, H., T. Kashimoto, and N. Kunita (1977) Detection and determination of polychlorodibenzofurans in normal human tissues and Kanemi rice oils caused 'Kanemi Yusho.' J. Food Hyg. Soc. 18(3):260-265.

Muto, T., T. Kabodera, and Y. Shinozaki (1974) The solubility of PCBs (polychlorinated biphenyls) in water and adsorption from water. Radiosotopes 23(4):19-22.

Nagasaki, H., S. Tomii, T. Mega, M. Marugami, and N. Ito (1972) Hepatocarcinogenicity of polychlorinated biphenyls in mice. Gann 63:805.

Nebeker, A.V., F.A. Puglisi, and D.L. DeFoe (1977) Toxicity of Polychlorinated Biphenyls (PCBs) to Fish and Other Aquatic Life. Report No. PB-264 804. Springfield, Va.: National Technical Information Service.

Nisbet, I.C.T. (1976) Criteria Document for PCBs. Report No. PB-225 397. Washington, D.C.: U.S. Department of Commerce.

O'Connors, H.B., Jr., C.F. Wurster, C.D. Powers, D.C. Biggs, and R.G. Rowland (1978) Polychlorinated biphenyls may alter marine trophic pathways by reducing phytoplankton size and production. Science 201:737-739.

Platonou, N.S. and B.S. Reinhart (1973) The effects of polychlorinated biphenyls (Aroclor 1254) on chicken egg production, fertility and hatchability. Can. J. Comp. Med. 37:341-346.

Powers, C.D., R.G. Rowland, H.B. O'Connors, Jr., and C.F. Wurster (1977) Response to polychlorinated biphenyls of marine phytoplankton isolates cultured under natural conditions. Appl. Environ. Microbiol. 34(6):760-764.

Sanborn, J.R., W.F. Childers, and R.L. Metcalf (1975) Uptake of three polychlorinated biphenyls, DDT, and DDE by the green sunfish, *Lepomis cyanellus* Raf. Bull. Environ. Contam. Toxicol. 13:209.

Shiota, K. (1976) Embryotoxic effects of polychlorinated biphenyls (Kaneclors 300 and 500) in rats. Okajimas Fol. Anat. Jap. 53:93-104.

Stalling, D.L. and F.L. Mayer, Jr. (1972) Toxicities of PCBs to fish and environmental residues. Environ. Health Perspect. 1:159-164.

Stotz, I.J. and Y.A. Greichus (1978) The effects of a polychlorinated biphenyl Aroclor 1254 on the white pelican: ultrastructure of hepatocytes. Bull. Environ. Contam. Toxicol. 19(3):319-325.

Tucker, E.S., W.J. Litschgi, and W.M. Mees (1975a) Migration of polychlorinated biphenyls in soil induced by percolating water. Bull. Environ. Contam. Toxicol. 13(1):86-93.

Tucker, E.S., V.W. Saeger, and O. Hicks (1975b) Activated sludge primary biodegradation of polychlorinated biphenyls. Bull. Environ. Contam. Toxicol. 14(6):705-713.

U.S. Environmental Protection Agency (1972) An Evaluation of DDT and Dieldrin in Lake Michigan. Ecol. Res. Ser. Report No. EPA R3-72-003. Washington, D.C.: Environmental Protection Agency.

Weil, L., G. Dure, and K.E. Quentin (1974) Water solubility of chlorinated hydrocarbon insecticides and PCBs in respect to water loading by these compounds. Wasser Abwasser Forschung 7:169-175.

Wong, P.T.S. and K.L.E. Kaiser (1975) Bacterial degradation of polychlorinated biphenyls II. Rate studies. Bull. Environ. Contam. Toxicol. 13(2):249-255.

Wyndham, C., J. Devenish, and S. Safe (1976) The *in vitro* metabolism, macromolecular binding and bacterial mutagenicity of 4-chlorobiphenyl, a model PCB substrate. Res. Commun. Chem. Pathol. Pharmacol. 15:563-570.

Zitko, V. and P.M.K. Choi (1971) PCB and Other Industrial Halogenated Hydrocarbons in the Environment. Technical Report No. 272. St. Andrews, New Brunswick, Canada: Fisheries Research Board of Canada.